Bergson's Philosophy
of Biology

Bergson's Philosophy of Biology

Virtuality, Tendency and Time

Tano S. Posteraro

EDINBURGH
University Press

Edinburgh University Press is one of the leading university presses in the UK. We publish academic books and journals in our selected subject areas across the humanities and social sciences, combining cutting-edge scholarship with high editorial and production values to produce academic works of lasting importance. For more information visit our website: edinburghuniversitypress.com

Edinburgh University Press Ltd
The Tun – Holyrood Road
12(2f) Jackson's Entry
Edinburgh EH8 8PJ

First published in hardback by Edinburgh University Press 2022

Typeset in 11/13 Bembo by
IDSUK (DataConnection) Ltd, and
printed and bound by CPI Group (UK) Ltd, Croydon, CR0 4YY

A CIP record for this book is available from the British Library

ISBN 978 1 4744 8880 8 (hardback)
ISBN 978 1 4744 8881 5 (paperback)
ISBN 978 1 4744 8882 2 (webready PDF)
ISBN 978 1 4744 8883 9 (epub)

Contents

Acknowledgements

I would like to thank Len for supervising most of the research for this book. I would like to thank Barry for sharing his own work on Bergson and helping me refine some of my formulations. I am indebted to Carol MacDonald and the rest of the team at Edinburgh University Press. Thanks to Ben for the late-stage feedback. Thanks to Laurence for everything else.

Abbreviations

C	Bergson, Henri. 2002. *Correspondences*. Paris: PUF.
CE	Bergson, Henri. 1998. *Creative Evolution*. Trans. Arthur Mitchell. Mineola: Dover Publications.
CM	Bergson, Henri. 2007. *The Creative Mind: An Introduction to Metaphysics*. Trans. Mabelle L. Andison. Mineola: Dover Publications.
HTM	Bergson, Henri 2018. *Histoire des théories de la mémoire. Cours au Collège de France 1903–1904*. Ed. Arnaud François. Paris: PUF.
L	Bergson, Henri. 2007. 'Cours sur le *De rerum originatione radicali* de Leibniz'. Ed. Matthias Vollet. *Annales bergsoniennes III: Bergson et la science*. Ed Frédéric Worms. Paris: PUF. 35–52.
M	Bergson, Henri. 1972. *Mélanges*. Ed. André Robinet. Paris: PUF.
ME	Bergson, Henri. 1920. *Mind-Energy*. Trans. H. Wildon Carr. New York: Henry Holt and Company.
MM	Bergson, Henri. 2004. *Matter and Memory*. Trans. Nancy Margaret Paul and W. Scott Palmer. Mineola: Dover Publications.
PL	Bergson, Henri. 2017. *L'évolution du problème de la liberté. Cours au Collège de France 1904-1905*. Ed. François Aarnaud. Paris: PUF.
TFW	Bergson, Henri. 2001. *Time and Free Will: An Essay on the Immediate Data of Consciousness*. Trans. F. L. Pogson. Mineola: Dover Publications.
TS	Bergson, Henri. 1977. *The Two Sources of Morality and Religion*. Trans. R. Ashley Audra and Cloudesley Brereton. Indiana: University of Notre Dame Press.

Introduction: Between Philosophy and Biology

This is a book on Henri Bergson as a philosopher of biology, for readers new to Bergson, Bergson scholars, and philosophers of science interested in but unfamiliar with his views.

While a new wave of enthusiasm for Bergson's philosophy has been rising in anglophone circles since at least the early 1990s, and while we now seem to be in the midst of what some have called a 'return to Bergson' and even a full-blown 'Bergson renaissance', a number of obstacles to Bergson's uptake as a serious philosopher of biology remain (see Guerlac 2006: x; Kreps 2015: 1; Ansell-Pearson 2018: 1; Lundy 2018a: 5; Sinclair 2020: 2). Among the most significant of these are his association with the legacy of vitalism, his commitment to a scientifically untenable metaphysics, and the progress made by research in the life sciences since (see Canales 2015: 7; cf. Quick 2017). As a result, few philosophers of science have engaged with his work – if they have read or heard of it at all.

This book paints a different portrait. I argue that the typical criticisms are mistaken, and that Bergson should be understood as a philosopher of biology with contemporary relevance. One part of my aim is to provide readers already familiar with Bergson with a systematic appreciation for his philosophy of biology. Another is to provide otherwise unfamiliar readers with a good sense of why Bergson should be taken seriously in relation to the history and philosophy of the life sciences. In the chapters that follow, I distinguish his views from those of traditional vitalism and provide a fresh interpretation of his metaphysics that answers to contemporary concerns. I also show how Bergson's engagement with the mechanistic and teleological assumptions in the sciences of his time can be updated according to current trends in adaptationism and gene-centric understandings of evolution and development. On my account, Bergson's thinking should be considered a resource for the recent turn towards process in the philosophy of biology, for the thermodynamic conception of organisation, for the relationship between structure and function, as well as for ongoing attempts to grasp the nature and significance of convergent evolution, the deep unity of life, the metaphysics and epistemology of evolutionary novelty, and a number of other issues in the philosophy of biology besides.[1] For these reasons, this book should appeal to those interested in the history and philosophy of biology and evolutionary theory in general, and in early twentieth-century philosophical responses to Darwinism in particular.

Bergson has long been considered a sophisticated resource for thinkers engaged by the significance of the natural sciences for philosophy (see Dobzhansky 1967). The initiation of a concern for Bergson's philosophy of life in particular is probably best marked by the publication of *The New Bergson* in 1999, which included an early article written by Keith Ansell-Pearson on the topic of organic life in *Creative Evolution* (1999). Ansell-Pearson's *Philosophy and the Adventure of the Virtual*, which appeared three years later, presented a portrait of Bergson sketched on the basis of Bergson's philosophy of life. Ansell-Pearson demonstrated both the extent to which the philosopher could be understood in this form as well as the extent to which Bergson's philosophy of life could be redeployed anew in the context of contemporary science (cf. Ansell-Pearson 2005b).

The centennial of *Creative Evolution* was marked in 2007 by the publication of a special issue of *SubStance* dedicated to the relationship between Bergson's philosophy of life and contemporary evolutionary theory (Kolkman and Vaughan 2007). Paul-Antoine Miquel's article on the convergences between Bergson and Darwin was distinctive for the rigorous study of Bergson's relationship to the evolutionary sciences (2007a). This style of scholarship was developed further in *The New Century*, a volume co-edited by Ansell-Pearson and Alan Schrift in 2010, the central chapter of which provides an exemplary study of Bergson's relevance for reflection on contemporary biology (Ansell-Pearson, Miquel, and Vaughan 2010). An historical supplement and counterpart was provided in the same year by Carvalho and Neves (2010), and then again more recently by Emily Herring (2018).

The renewed appreciation for Bergson's philosophy has resulted in a twenty-volume critical edition of his work and the publication of his Collège de France lecture courses. New scholarship has begun to recognise his pivotal contribution to the history of modern philosophy as well as his ongoing relevance to the issues animating a number of areas of philosophy today. At the same time, there has been an efflorescence of philosophical interest in the life sciences, which seem to be at an especially productive moment of development. There are opportunities emerging for the philosopher of these sciences for speculation regarding the novel directions they might begin to take, as well as for reconsidering the earlier critics of their current formation. There can be found in Bergson a set of resources for both tasks. He was an ardent critic of what came to be called Neo-Darwinism, especially concerning the commitment to pan-adaptationism and the idea that natural selection could provide an exhaustive explanation for evolutionary creativity, directional variation, and convergence. While his work has been out of fashion for many decades, the problems that animated it and to which it was originally addressed remain pressing for the philosophy, the biology, and the philosophy of biology of today.

This is a book written across these two contexts. The ambition is for it to mark an advance in the increasingly serious, critical, and contemporary study of Bergson's work. What distinguishes this book from the other recent monographs on Bergson is the focus on his philosophy of biology, and its engagement with the contemporary life sciences in that connection.[2]

Vitalism, psychology, and metaphysics

Bergson has been criticised for his vitalism, for privileging first-person conscious experience in the interpretation of biological phenomena, and for his apparent irrationalism and his metaphysics. These criticisms are rooted in a set of common interpretive errors. The first is to misconceive Bergson's *élan vital* as a vital principle of the traditional variety, instead of as a heuristic 'image' for the dynamic nature of evolution. The second is to subordinate Bergson's philosophy of biology to his analysis of lived experience, which misrepresents the pragmatic character of what Bergson calls the 'psychological interpretation' of biological phenomena. The third is to conflate Bergson's rejection of rationalist metaphysical systems with a rejection of philosophical systematicity *tout court*. In this book I identify each as an error and explain why they should be avoided. Correcting them allows me to elaborate a largely novel, systematic interpretation of the main themes of Bergson's philosophy of biology, rendering him more plausible as a philosopher of biology as a result.

Vitalism. Vitalism is typically taken to consist in the contention that matter and the mechanical models that track it are insufficient to account for the reality of biological forms, and that the explanation of life therefore requires the postulation of a non-mechanical, possibly immaterial, uniquely vital principle, force, substance, or property.

Bergson's infamous *élan vital* is commonly regarded as one such principle. It is usually considered an object of possibly historical interest, but no longer a seriously credible idea (see Azouvi 2008). To the extent that *Creative Evolution* revolves around a metaphysics of vital force, it is badly outdated, having nothing to say to contemporary thinking on any of its topics. Advances made in the life sciences since Bergson's time have rendered his positions untenable. His criticisms no longer track current theories, and his positive postulates have been undermined by the research. *Creative Evolution* may be an important work of early twentieth-century philosophy, but as a work in the philosophy of biology it is obsolete. Though some recent scholarship has attempted to complicate Bergson's status as a vitalist, it either has not yet gone far enough in differentiating the *élan vital* from a vital principle or force, has not presented the arguments requisite for the conclusion that Bergson is not a vitalist, has conceded too much to vitalist readings of Bergson in the process, or else has stopped short of a complete rereading of Bergson's philosophy of biology.

Bergson's Philosophy of Biology aims to combat this received view in each of its assumptions. In this respect, I follow the lead of scholars like Jean Gayon (see 1998, 2005, 2008), who complicate the picture of a scientific enterprise either indifferent or hostile to Bergson's philosophy on the one hand, and an interpretation of Bergson's philosophy more or less irrelevant to the history and philosophy of the life sciences on the other. To signal this, I call Bergson a philosopher of biology, not a philosopher of life, as is the custom. Partly this is in order to distinguish him from the figures of Romantic Biology and *Lebensphilosophie*. Partly it is in appreciation of the fact that many of his views are worked out in contact with the extant biology of the late nineteenth and early twentieth centuries. Most importantly, I call Bergson a philosopher of biology because I do not believe that he produces or defends a concept, principle, or force that can be usefully nominated under the aegis of Life. Contrary to one of the distinguishing marks of his reputation, I argue that Bergson is not a vitalist in the way that the designation has been traditionally understood.

If there is something that sounds like a vital principle in Bergson's thinking, it is obviously the *élan vital*. In this book, I approach the *élan vital* not as a concept with an identifiable object, like a vital force, but as an image intended to point biology beyond the inadequacy and reductionism of the concepts already available. Life – organisation and evolution – is supposed to elude the alternative between efficient and final causality – mechanism and finalism – which means that it cannot be determined in the causal terms of either. In Chapter 1, I show how mechanist and finalist forms of biological explanation deprive evolution of the novelty that Bergson considered to be one of its ineliminable features. Bergson thought for this reason that life could be better approached through an 'image', an image of an impetus or an '*élan*' in particular.

Images serve as tools for the expression and communication of thoughts that frustrate or escape the delimited boundaries of available concepts. Images point to something outside of those concepts. In this sense, the image of the *élan vital* serves a heuristic function, regulating against the tendency to reduce the dynamic nature of organisation and evolution to what can be captured in mechanistic models, explained in terms of adaptation, or foreseen in the form of a genetic program or pre-existent possibility. Positively, the image of the *élan* indicates that there is something more at work in patterning the evolutionary movement. This 'something more' is what Bergson calls tendency. The image of a vital *élan* is intended to indicate an idea of tendency that remained underde-veloped in Bergson's work. The prosecution of this argument regarding tendency is one of the features that makes my interpretation distinctive.

The main interpretive move of the book is to make sense of the idea of tendency as Bergson employed it in *Creative Evolution* in terms of the idea of virtual existence that he elaborated in *Matter and Memory*. The modal status of tendency, its mode of existence, is virtual, like that of unrecalled memory. The idea of tendency is not that of potential-ity, an idea familiar to the history of philosophy. The virtual existence of tendencies consists in the non-teleological inclination to actuality. Tendencies tend to become all that they can be. Individual tenden-cies contract the interpenetrating whole from which they are dissociated as they develop in increasingly particular directions over time, just as distinct memories contract the whole past that they make present in a recollection. The idea is to make sense of Bergson's contention that 'life is tendency' on this model.

This allows me to situate Bergson's idea of tendency against the Darwinian view that chance variations producing a random pool of indi-vidual differences furnish the conditions requisite for natural selection – the principal cause of evolution – to do its work. Bergson agreed that natural selection works on variations, which are individual differences, but denied that they are chance or accidental occurrences, what today we understand as mutations and copying errors. Life, for Bergson, does not evolve through a series of accidents. It is trended by tendencies, and patterned by their materialisation. As a result, natural selection should not be considered the omnipotent evolutionary agency that Neo-Darwinians assume it to be. It can only be so omnipotent if variations are random, playing no efficacious role in the directions evolution takes. That is what Bergson contested. He thought convergent evolution made his point.

I use the idea of tendency to reconstruct his argument and consider it against the current science. Convergent evolution refers to the repeated appearance of like characters, such as the camera-like eye, across distant lineages like snails and spiders, or humans and octopuses, evolving in dissimilar environments. It is an infamously difficult phenomenon to explain by dint of natural selection for chance variations alone. Bergson's view is that convergence is not an accident or anomaly of an otherwise random variational process, but that it rather reveals something important about how evolution works. What it reveals is a deep unity – or what researchers call homology – underlying the contingent plurality of life forms, and a pattern or trend to their repeated convergence on the same organs and traits.

Evolution, on my reconstruction of Bergson's position, is trended by tendencies, acquiring a directional patterning as it unfolds. Selection acts on directional potential, shaping a set of tendencies in addition to a collection of traits. Those tendencies contract the unified whole from out of which they are progressively dissociated over evolutionary time. Convergent traits are less the result of two lineages accidentally happening upon the same solution in external concord and more a trace of the deep history in which all tendencies are ultimately unified. The contemporary study of conservation, deep homology, and developmental constraint all point in the direction of a patterned and trended understanding of the evolutionary process. Considered in this connection, Bergson's idea of *élan vital* is closer to the contemporary science than to the history of vital principles.

In reconceiving the *élan vital* as an image for tendency, the goal is to distinguish Bergson's philosophy of biology from traditional vitalism without doing away with its metaphysical character. Retaining that metaphysics also allows me to update Bergson's criticisms of mechanism and finalism in the biology of his time. Recent gene-centric understandings of evolution and development are not free from metaphysical assumptions. In many cases, they continue to operate on the same assumptions that Bergson identified (see Goddard 2010). Those assumptions involve a conception of the organism as an organised set of parts selected for their adaptation to environmental pressure, a conception of development that sees in the organism a product of the program of its genes, and a conception of evolution that sees in the emergence of new structures the realisation of pre-existent possibilities mappable in advance. In each instance it is the reality of time that is at stake. Bergson's criticisms of biology's spatialising tendencies continue to track reductive trends in the life sciences.

Bergson may have agreed with Joseph Woodger that biologists who 'suppose themselves to be above "metaphysics" are only a very little above it – being up to the neck in it' (1929: 246). He may have also agreed with Whitehead that 'every scientific man in order to preserve his reputation has to say he dislikes metaphysics. What he means is he dislikes having his metaphysics criticized' (qtd. in Conger 1927). Referring to the recent publication of *Creative Evolution*, Bergson told Hans Driesch that 'if a book such as mine can contribute to eliminating the unconscious (and hence inconsistent) metaphysics that permeates a good deal of our evolutionism, I would be truly happy' (C 160). The relevance of *Creative Evolution* is to be found not only in its positive accounts of the unity, novelty, and directedness of the evolutionary process, but also in its critical engagement with the metaphysics at work behind reductive models of development, mechanistic descriptions of organisation, and pan-adaptationist explanations of evolutionary events as well.

Psychology. Bergson began his career with the analysis of lived experience, taking personal consciousness as a privileged domain in which to explore topics such as time, space, memory, creativity, change, individuality, and freedom. He proceeded to attribute psychological characteristics to the evolutionary process, using consciousness as a synonym for what the sciences could not grasp about evolution and arguing on behalf of a 'psychological interpretation' of nature (CE 257). The troubling implication was that either Bergson believed there to be something like a transcendent 'intelligence' of nature responsible for organising and directing it, or that he subordinated the empirical research of the life sciences to the first-person data of introspective analysis.

That reading misconceives both the status and aim of what Bergson calls the psychological interpretation. I claim that the status of Bergson's interpretation is heuristic, not ontological, and that its aim is not to provide an exhaustive account of nature, but to serve as a means by which to grasp the irreducibly temporal features of biological phenomena. These features include the preservation of the past in organic memory and evolutionary history; the developmental continuity and interpenetration of parts in an organised whole; and the unforeseeable elaboration of new forms in evolutionary time. The idea is that psychological terms are the best we have available for the project of describing these features, because we can learn to grasp the subjective instantiations of those features in ourselves first. The lived experience of psychological duration serves as a point of departure for the project of thinking about evolution's temporality in general.

Though Bergson begins with first-person experience, his mature view is that our experience is largely structured by evolutionary demands and tuned to adaptive action. Negatively, this means that the philosophy of biology requires a critical account of the genesis of intelligence. Positively, this means that the data of first-person experience can serve as a window onto the evolutionary dynamics that generated and inform it (CE xix–xx).

In his Introduction to the English translation of *Matter and Memory*, Bergson draws two principles from the first idea, that consciousness testifies to an evolutionary history.

The first is that we should not forget the utilitarian character of our mental functions, which are essentially turned towards action. The second is that the habits formed in action find their way up to the sphere of speculation, where they create fictitious problems, and that metaphysics must begin by dispersing this artificial obscurity (MM xx–xxi).

The action orientation of our cognitive capacities is adaptive, a function of our biology. The cognitive habits and heuristics that come naturally to us do so because of their practical efficacy. Intelligence evolved to serve the survival of the species, not for theoretical contemplation or disinterested knowledge (see Ruse 2002). Appreciating that means being able to identify false or fictitious problems, problems posed because of the misuse of utilitarian intellectual tendencies outside of the domain of action. Many of the perennial issues of philosophy are problems of this sort. Any philosophy that ignores the biological basis of the human intellect risks mistaking an unconscious heuristic for an accomplishment of philosophical knowledge. The critique of false problems, a theme familiar to commentaries on Bergson, relies on an understanding of evolution as the conditioning cause for the formation of the intellect in the first place.

The second idea is that the intellect, since it is a product of evolution, can be employed as a means by which to construct a philosophy of evolution as well. Theories of mind, knowledge, and experience are to be grounded in a theory of life. This is because

> a theory of knowledge which does not replace the intellect in the general evolution of life will teach us neither how the frames of knowledge have been constructed nor how we can enlarge or go beyond them. It is necessary that these two inquiries, theory of knowledge and theory of life, should join each other, and, by a circular process, push each other on unceasingly. (CE xxiv)

In this book, I situate Bergson's analysis of lived experience within his analysis of evolution. If lived experience is an adequate point of departure for philosophical reflection on nature, it is because the dynamics of conscious life contract and recapitulate the dynamics of evolution on a local scale. Structured through the movement of evolution, they testify to how that movement unfolds. The inquiry into consciousness is a means by which to pursue an investigation into the structure of nature.

Whereas Kant endorsed an inviolable distinction between our phenomenal experience of the world and the world as it is in itself, Bergson denies that there is a difference in kind between perception and reality. There is, on his view, only a difference in degree between phenomenal experience and the noumenal world. That difference has its genesis in evolution, as our perceptual systems evolve according to what it is useful for us to perceive in our environments. Perceptual experience is an index, in part, of that evolutionary adaptation. Approaching the issue from this perspective, I show how Bergson begins with lived experience in order to get beyond it, providing a philosophy of nature that subjective experience instantiates.

Bergson's psychological analysis, on my account, is an investigation into nature subjectivised. His study of nature proceeds on the basis of his psychology. In this respect, I follow Gayon's observation that the study of evolution provided Bergson with 'the empirical material capable of offering the largest extension of his indeterminist theses and his vision of mind' (2008: 61–2). I add that knowing more about nature means knowing more about knowledge in turn. The consequence is a self-correcting reciprocity between biological research and philosophical speculation, with no end to the recursion foreseeable in advance.

In a lecture from 1903, Bergson contended that 'philosophy has some value only on condition of being always, constantly verified by contact with the positive sciences' (HTM 26). In 1911, he implored philosophers to

> study the ancients, become imbued with their spirit and try to do, as far as possible, what they themselves would be doing were they living among us. Endowed with our knowledge (I do not refer so much to our mathematics and physics, which would perhaps not radically alter their way of thinking, but especially our biology and psychology), they would arrive at very different results from those they obtained. (CM 107–8)

I take these statements as guiding lights throughout this book. I locate Bergson accordingly – between philosophy, biology, and the philosophy of biology. *Bergson's Philosophy of Biology* refracts each through the others in the pursuit of a biologically informed philosophical system, a biological worldview enlarged by the philosophy that is its correlate, and a philosophy of biology adequate to the latter by virtue of being formed through the unfolding of the former.

Metaphysics. This book aims to show that Bergson's philosophy, his philosophy of evolution in particular, can be reconstructed as a work in metaphysics, in something like a systematic fashion. The common rejection of this interpretive possibility has fed the misconception that Bergson is an irrationalist, denouncing scientific rationality and mathematical intelligence in favour of an unmediated, mystical access to reality called intuition (see DiFrisco 2015: 56).

Bergson sometimes indicated that each of his books and the sets of concepts they developed was meant to address a particular constellation of problems (CM 87–106, 170–6; cf. During 2004 and Lundy 2018b). As the problems changed, so did the concepts. Even if it continued to go by the same or a similar name, each concept was to be cut anew according to the size of each new object, made to fit as closely as possible in every case. It might be said that the virtual of *Matter and Memory* differs from the same idea as it is employed throughout *Creative Evolution*. In the former text, virtuality is a term for the existence of memory in the past. In the latter, it is a term for tendencies alive with directional potential. Reading across each work and synthesising the ideas and arguments produced in response to one set of problems with those produced in response to another might seem to require too much abstraction. Bergson did not present his own philosophy in systematic form. Why do it for him?

There is a difference between the idea of absolute systematicity that Bergson rejected and the systematic form that I give to his views here. For one, an absolute system is supposed to remain insulated against progress made by the sciences. It is defined in part by a total conception of nature that is not amenable to new discovery. For another, it is supposed to be derived rationally, from axiomatic postulates. Empirical data might be employed to bear it out, but cannot be used to inform or falsify it in any serious sense. The classical philosophical system cannot be updated or concretised. It is an invariant whole, to be taken or left as it is.

Bergson was an ardent critic of rationalism. His criticisms of systematic philosophy were criticisms of philosophical rationalism. His own thinking does not tend away from systematic consistency but away from rationalist presuppositions. His philosophy develops a number of themes distinctive of philosophical empiricism. Like all empiricists, Bergson stresses the pragmatic orientation of intelligence and the utility of our concepts. Intelligence is Bergson's term for mechanistic thinking, calculative rationality, geometric reasoning, causal determinism, counting, measurement, reduction. His anti-intellectualism is a stance against the idea that these cognitive and conceptual strategies tell us anything about nature as it is. Intellect evolved in the way that it did, with the forms and faculties that it has, because it helps our species act on relevant features of our environment. Rationalism, with its pretence to absolute knowledge on the basis of reason alone, mistakes a contingently evolved set of adaptive habits for unmediated access to the truth.

Science should be understood in the same way. It is at best a partial view on nature, oriented towards extending our action, our ability to make, manipulate, and control. For that it employs a set of spatialising assumptions that make nature responsive to experimental inquiry. Scientific knowledge does not go much further than that. At its most mechanistic, it eliminates everything that makes experience what it is, qualitatively heterogeneous, indivisibly continuous, durational. The mistake is to think that nature itself must be matter in motion and nothing besides and that duration must be merely subjective, an artifact of psychology, not a real feature of nature. What Bergson says about rationalism he says about science as well. Experience is experience in time. There is no experience without duration. If scientific mechanism dispenses with the unforeseeable novelty that is definitive of time, then it is scientific mechanism, not time, that has to be overcome.

Bergson did not deny that philosophy could take the form of a system, only that system could be understood as rationalism understands it. He suggested that philosophical systems could be better conceived on an analogy with the structure of organisms (CM 91–103). The activity through which philosophical systems are constructed is analogous to the process through which life evolves and self-organises, recapitulated in another register. Conceived in terms of their composition, organisms are complex wholes that integrate a plurality of parts at a number of levels. Conceived in terms of the activity through which they come to be composed, organisms are simple and continuous processes of differentiation. Their complexity is derivative on the simplicity of the developmental movement. Understood in this way, 'a philosophy resembles an organism rather than an assemblage' (CM 91). It is not assembled out of discrete parts. Its parts are the products of the continuous process of their formation. The philosophical system appears complex only from the point of view of its conceptual architecture, which develops around 'the novelty and simplicity of its inner content' (CM 91). Its exterior articulations are like the multiple parts of a living thing; its inner content is like the activity of the development of its parts. At the centre of a philosophical system is not a set of static postulates, but an impulse or movement. The process of explication involves the extension of an intensive insight across a series of terms, concepts, contentions, arguments, and expressions.

Bergson rejected the classical ideal of systematic unity because philosophy, like life, does not culminate in a state of unity but rather develops from out of one. The original unity of thinking is 'a unity which is at once restricted and relative, like the unity which marks off a living being from the rest of the universe' (CM 103). Just as the organism develops in contact with its milieu, maintaining itself in a state of dynamic stability by drawing from what is outside it, the philosophical system should develop in exchange with what is outside it as well, the evolving scientific data from which it draws and upon which it exerts its own influence in turn. At its intensive centre, philosophy is metaphysics; at its outer rings, it is open empirically to innovations in the scientific determination of the natural world (CM 162). The developmental interface between the two is what constitutes the 'true' or 'superior' empiricism for which Bergson first argued in 1903, and in which *Creative Evolution* is intended to culminate (CM 147, 102–3). These are the ideas behind my system-like presentation of Bergson's thought.

The contention that there is a set of concepts that Bergson's work develops as a whole, and that those concepts can be isolated from each of the contexts in which Bergson first formulated them, is most readily associated with the Deleuzian interpretation (see Lundy 2018a: 1–16 and Bianco 2011: 858). Like many contemporary readings of Bergson, my account has been influenced by Deleuze's in many of the key details. That said, the concepts that I emphasise are, with the exception of multiplicity and virtuality, not the same ones that Deleuze made popular in his work. I accord a central significance to the ideas of image, tendency, individuality, and finalism, as well as canalisation and convergence in the evolutionary domain.

I try to show that Bergson developed a set of metaphysical views over the course of his thinking, and that these views constitute something like a unified philosophical system. They can be presented argumentatively, in ways that bear out their ongoing relevance. To some extent, they can be disarticulated from their original formulations and brought into contact with contemporary scientific research. Given the heuristic function of concepts, I do not think that they have any claim to the truth. They are not exhaustive of their objects, totalising of all biological phenomena, or foundational for any attempt to make sense of nature. Yet they continue to offer resources for the project of making sense of life today.

What of the idea that the concepts of one text are not those of another? Just as there are two kinds of system, there two kinds of concept as well. Most concepts are 'fixed', 'rigidly defined', 'inflexible and ready-made' (CM 160). These are the concepts that come naturally to the intellect, tuned as it is to practical action. When they are imported into philosophical speculation, they spatialise, stabilise, fix, and decompose the dynamic nature of their objects. They erase duration, decompose wholes, isolate and localise parts, make things calculable and predictable. The critique of intelligence is a critique of concepts of the ready-made sort. It would be a mistake to try to capture Bergson's philosophy with these concepts or the categories and positions with which they have been associated, like mechanism or finalism, realism or idealism, spiritualism or materialism, emergentism or panpsychism, and so on. Although this can be an illuminating exercise, it fails to account for what is novel about the Bergsonian project.

Not all concepts are ready-made. Bergson's texts are populated with concepts of a different kind. They are concepts that are formed in the act of thinking a reality that frustrates and escapes the delimitations of concepts already available. After moving beyond ready-made concepts, 'the new concepts one must form in order to express oneself will now be cut to the exact measure of the object' (CM 17). These concepts are 'fluid, capable of following reality in all its windings and of adopting the very movement of the inner life of things' (CM 160). Their articulation is one of the aims of philosophical thought. The concept of tendency is one example.

The issue is how to know when a fluid concept can be employed appropriately in order to track the movement of some object, and when a new concept must be formed according to the measure of the new object instead (CM 17). Fluid concepts are always at risk of congealing, becoming fixed, applied too rigidly in a way that effaces the dynamic nature of new objects. Extracting concepts from the contexts of their formation risks rigidifying them, but it does not necessitate that consequence. When Bergson draws the concept of qualitative multiplicity from *Time and Free Will* into *Creative Evolution*, he is developing the concept further, deepening its scope, not fixing and emptying it. Whether this book immobilises Bergson's concepts or manages to maintain their fluidity by extending them across new research and reinvigorating them beyond their original contexts is a question for its readers.

Outline and overview

I have structured the book as a series of interpretations of the central topics in Bergson's philosophy of biology. I think this is the best way to render his thinking plausible to those unfamiliar with his work while still developing my own interpretation of it. This is the best way to make Bergson's philosophy of biology plausible, as it allows me to lay out the distinctive features of his views on time, space, change, process, tendency, unity, multiplicity, organisation, and so on, without assuming too much background knowledge as I proceed. My goal is to provide only as much textual exegesis as is necessary to understand how Bergson makes a contribution to thinking about each issue.

I have structured the book in this way for systematic reasons as well. My position is that Bergson's views on biology make up the fundamental components of his philosophical system. As part of the aim of this book is to build that system out, it makes sense to do so by proceeding through one topic at time. In this way, the book makes it clear for philosophers of science what sort of philosophy of biology Bergson's thinking entails, while also showing, for Bergson scholars, how his views on familiar themes should be understood on the basis of his philosophy of biology.

The book is divided into six chapters. The first chapter begins with Bergson's criticisms of the scientific conceptions of life and explanations for evolutionary convergence of his time. I reconstruct them on metaphysical grounds, showing how they take for granted assumptions that negate the reality of time and the novelty of evolution. In the second chapter I elaborate Bergson's alternative metaphysics, that of the virtual, in terms of virtual wholes that are contracted in each of their parts. In Chapter 3, I use this theory of the virtual to develop my interpretation of a concept of tendency that Bergson worked out across each of his texts. Tendency is directional inclination. Its modal status is virtual. Tendencies are qualitative multiplicities, virtual wholes tending into concrete actuality through processes of bifurcation, dissociation, and the determination of parts. In Chapter 4, I show how this idea provides the key to a number of important concepts and distinctions in Bergson's thinking. I argue that tendency is at work beneath the time/space and life/matter dyads, and that the *élan vital* is best understood in terms of tendency as well. Chapter 5 turns from individual organisms to the evolutionary scale, and situates Bergson's idea of tendency in the context of the theory of orthogenesis, or directional evolution. I argue for a heuristic reading of Bergson's 'psychological interpretation' of orthogenesis and present my view that the status of the *élan vital* is an image instead of a force or principle. Chapter 6 completes the book by returning to the case of convergent evolution introduced at the beginning and showing how the metaphysics of tendency that the book develops provides an account that is consonant with the contemporary science on convergence.

*　*　*

The goal of the first chapter is to isolate and reassemble the critical moments of Bergson's philosophy of biology in argumentative form, showing what it was that he was arguing against. I represent these criticisms as arguments against the metaphysics of the 'actual' and the 'possible' in mechanist and finalist biology. Actuality and possibility are modal categories, types of existence. Mechanism relies on a metaphysics of actuality to the extent that it regards determinate organic parts as exhaustive of biological phenomena. Adaptationism is evolutionary mechanism. This form of thinking negates the reality of time by isolating the organism's parts and processes in the present and placing them in a passive causal relation with selection pressures from the outside, on the model of an unorganised material system. Finalism relies on a metaphysics of possibility to the extent that it assumes that developmental ends or evolutionary events can be given at the outset of the processes that lead to them. This form of thinking negates the reality of time by determining the future in advance, in the form of possibility, rendered in terms of developmental plans, genetic programs, blueprints, or functional ends, thinkable in advance.

There are, however, two forms of finalism, internal and external, one that locates purposiveness internal to the organised wholes of individual organisms and one that locates purposiveness external to them, in the ends that orient their development or the possibilities that they realise over the course of evolution. I reconstruct Bergson's arguments against both, but spend more time developing the criticism of internal finality, what Kant called inner purposiveness. My presentation of that argument sets the stage for the book's interpretation of evolutionary tendency as a form of external finalism by showing how the internal view mistakenly regards organisms as closed individuals. Bergson's position is that biological individuals are only ever relatively closed, such that if there is to be finality in the organic domain, it will have to be rendered external to all ostensible individuals. Appreciating that point enables me to clarify an important part of the motivation for Bergson's view of evolutionary tendency, that is, that there is more to life than the organisation of organisms. I conclude the chapter by tracing the arguments against mechanism and finalism through some of their contemporary iterations in gene-based understandings of development and evolution. This chapter will appeal to philosophers of biology concerned with reductionism in the sciences of evolution and development, as well as those who are curious about the metaphysical assumptions that those sciences take for granted.

Having shown in Chapter 1 that Bergson considers mechanist and finalist biological explanations to be contaminated by metaphysical assumptions regarding actuality and possibility, I turn in Chapter 2 to

Bergson's alternative modal category, that of the virtual. The idea is that if mechanism and finalism both fail to capture the dynamic nature of evolution, then a different metaphysical modality is required. This chapter initiates the book's development of that metaphysics by drawing a conception of virtual existence from *Matter and Memory*. I show that the virtual refers throughout this text to wholes that tend towards the actuality of parts while both structuring and exceeding them. The whole is in each case always open, in a dynamic and coexistent relation with the parts into which it is dissociated, in which it is contracted, and in response to which it changes.

I extract this idea of an open whole that is instantiated in each of its parts from Bergson's discussions of the relationship between perception and environment, motor memory and its action schemas, and the pure past in relation to the particular memories that actualise it. The rest of the book leverages this modal mereology in service of reconstructing the critical and constructive dimensions of Bergson's philosophy of biology. I use the idea of virtuality to provide a systematic reconception of Bergson's views on tendency, creativity, the structure/function distinction in biology, convergent evolution and deep homology, the source and directionality of variation, the dynamic nature of metabolism, and a number of other crucial topics besides.

My view that there is a unified mereological theory of the virtual as a distinct modal category makes a contribution to the scholarship on Bergson and should be especially interesting for scholars concerned with the way virtuality is employed throughout his work. It should also interest readers of Gilles Deleuze, as Deleuze inherits his own theory of the virtual from Bergson. Philosophers of science engaged by discussions of actuality and possibility will find an alternative metaphysical framework for thinking about issues like unactualised capacities, dispositions, genetic information, evolutionary novelty, and organic memory as well.

Having presented the virtual as a general modal category in Chapter 2, Chapter 3 employs it in reinterpreting one of the central ideas of Bergson's philosophy of biology, the idea of tendency. This idea is still only beginning to attract the attention it deserves in the scholarship. Matthias Vollet (2008, 2012) has written two overviews of Bergson's use of the idea across his work, and Mark Sinclair (2020) has recently suggested that it is essential to understanding Bergson's theory of life. Bergson defines life in terms of tendency in *Creative Evolution*, and discusses the idea most explicitly in that connection. Though the relation of tendency to Bergson's theory of the virtual has been noticed, the two ideas have not yet been read together at length. I argue in this chapter that the theory of the virtual is the key to the idea of tendency, which is one of the main interpretive moves of the book; on that basis, I provide what is the most comprehensive account of tendency in the scholarship to date.

Chapter 3 begins by situating the idea of tendency against the background of Bergson's study of psychological effort. I show that the concept of tendency is meant to capture a form of effort de-subjectivised, dislocated from personal psychology and generalised across all directional activities. This sets the stage for the book's reinterpretation of the *élan vital* as an image for tendency drawn from psychological analysis. Next, the chapter traces the development of the idea of tendency across Bergson's work, from its initial employment in *Time and Free Will*'s introspective account of free acts through to its formulation as a concept in its own right in Bergson's last texts. My claim is that the status of tendency is virtual, and that it should be interpreted according to the mereological features of virtuality that I draw out in Chapter 2. By setting tendency apart from more familiar modal categories such as possibility and potentiality, I show how tendency accounts for actualisation processes while preserving the novelty of duration. This distinguishes Bergson's view from other theories of tendency on offer in the metaphysics of powers, for instance. I conclude the chapter by considering one recent example, from Anjum and Mumford (2011), and arguing that Bergson's conception avoids some issues that persist through the contemporary account. While the first sections of the chapter should be of interest to Bergson scholars, the last section should appeal to philosophers of science interested in ideas of causality, disposition, powers, and potentiality in nature.

Chapter 4 moves to reconsider Bergson's views on space and time on the basis of the metaphysics of tendency elaborated in Chapter 3. I use the idea of tendency to show how space and time should be understood as reified intellectual abstractions from spatialisation and temporalisation processes that characterise organised and unorganised material systems to different degrees. My claim is that intellect seizes on these processes, extrapolates them to the limit of each tendency, and then abstracts that ideal limit as a state or category of its own. The idea of inert material parts existing in a spatial container is not a mere mental projection or mistake, but an abstraction from the way matter behaves, tending towards spatial extension. Likewise, the idea of lived duration as a purely interior psychological phenomenon is an abstraction as well, as the features of duration characterise all organised matter to some extent. The scholarly contribution made by this interpretive move is to show how the genesis of the psychological concepts of Bergson's early work is to be found in Bergson's philosophy of biology.

It follows that the life/matter dualism that seems to structure much of the popular reception of Bergson's philosophy has been misunderstood as well. I argue that life and matter are not categories under which any actual entities could fall, any more than space and time are. This means that there is no such thing as inert matter on Bergson's view, and any reading of his thought that assumes a conception of inert matter imbued with *élan vital* has to be mistaken. Life and matter are rather abstract principles that define a set of conflicting tendencies towards time and space. All that exist in fact are composites formed out of the configuration of these tendencies under differing relationships of predominance and subordination. Organised material systems, or biological individuals, are in a particular sort of dynamic equilibrium between these two tendencies. Organisation is not a property, as the vitalists have claimed, but a manifestation of the tendency towards time achieving predominance in a material system. Organisms, in other words, are not so much types of entities as they are dynamic systems composed of interacting spatialisation and temporalisation processes.

The result of this interpretation is a renewed, processual account of what are often presented as the primitive terms of Bergson's philosophy, time and space, or life and matter. By understanding matter and life in terms of compositions of tendencies towards and away from temporal novelty, I am able to bring Bergson into dialogue with a number of more recent research programmes in the life sciences, such as Developmental Systems Theory and Nonequilibrium Thermodynamics, each of which complicates the life/matter dualism in productive fashion. Philosophers of science interested in the recent turn towards process in the philosophy of biology, as well as those interested in the philosophical dimensions of Developmental Systems Theory, should find these discussions particularly engaging.

Chapter 5 builds on Chapter 4's discussions of organisation and development by extrapolating the idea of tendency to the evolutionary scale. I show how thinking about tendency on this scale is the best way to understand the *élan vital*. My claim is that the *élan vital* is Bergson's image for a single, unified tendency, unfolding its directional progress through the medium of particular organisms. This position allows me to fill out the remaining details of Bergson's variation on external finalism, which began with his critique of the position that organisms are bounded individuals or organised wholes. Organisms are more like relatively stable, temporary formations that punctuate an evolutionary movement that flows through them. This is why finalism in the domain of life has to be rendered external to any ostensibly distinct biological individual. Explaining that idea marks a contribution to the secondary literature on Bergson's critique of Kant's theory of inner purposiveness by showing how that critique is completed by a positive view of purposiveness at a higher register.

By elaborating Bergson's external finalism of tendency, I am finally able to distinguish the *élan vital*, as an image for this tendency, from vital principles of the traditional variety. Those vital principles have typically been posited to secure the autonomy and irreducible individuality of organisation, as in Hans Driesch's embryology. Bergson's *élan vital* is an image for the evolutionary movement as a single event. Organisation is not a fact in need of a principle that would explain it, but a temporary form of stability composed of processes interacting at a number of time scales, taking place within the global movement of evolution as an open whole.

My account begins with a reconsideration of Bergson's reading of Weismann's germ theory, and concludes by revisiting Bergson's critique of biological finalism. After distinguishing the *élan vital* from traditional vitalism and conceiving it as a form of finalism modified to circumvent the issues Bergson identifies in other finalist accounts, I conclude with a discussion of what Bergson called the 'psychological interpretation' of evolution. My reading, which takes a pragmatic approach to the psychological nature of life, is what distinguishes this chapter from most other accounts of Bergson's purported vitalism as well as criticisms that target the apparent privilege he affords to psychology and the lived experience of duration. I argue that the psychological interpretation is a heuristic one, aimed at mobilising the experience of time in order to grasp the temporal nature of evolution without reducing life to mechanistic models or predetermining it according to finalist presuppositions.

In addition to scholars working on Bergson and the spiritualist tradition in French philosophy, my rendering of Bergson's psychological interpretation of evolution should appeal to those interested in what role there can be for introspection and lived experience in the construction of a philosophy of nature, such as scholars of German Idealism and American Pragmatism. For philosophers of science sceptical of the psychological interpretation, I provide a reading of Bergson that dispenses with the usual commonplaces regarding his subordination of empirical research to subjective experience. Philosophers of biology should find Bergson's anti-reductionist commitment to accounting for the dynamism, novelty, and non-mechanical directionality of the evolutionary process particularly appealing as well.

Chapter 6 concludes the analysis of Bergson's theory of tendency by returning to the case study of evolutionary convergence discussed in Chapter 1. I explain there that convergence – the repeated appearance of like traits across divergent lineages – requires a non-random understanding of the evolutionary process. Having demonstrated in Chapter 5 that the *élan vital* refers to a unified tendency unfolding on the evolutionary scale, I show in Chapter 6 how this account allows for a reconstruction of Bergson's answer to convergence in a way that accommodates for current research. An integral part of that reconstruction is Bergson's use of the image of a canal in his explanation of vision. I uncover a subterranean line of influence running from Bergson's use of this image, through the process philosophy of A. N. Whitehead, and into C. H. Waddington's embryological theory of how fields of possibility are progressively narrowed over the course of development. In tracing this line, I uncover a Bergsonian contribution to the structure/function distinction in the philosophy of biology. I show how, according to Bergson, function should be understood as enjoying a metaphysical privilege over structure, organic structures being in some sense the materialisation of the functions that they facilitate.

On my reconstruction of Bergson's view, evolution is 'canalised' or trended in the direction of convergent traits, such as complex eyes, because it is canalised in the direction of indetermination. Vision is one powerful way that indetermination is maximised, as the sensory presentation of choice to locomoting animals. The idea is that the function of vision is canalised across structures of varying complexity, visual organs. Where it attains to similar levels of sophistication, it is materialised in organs of similar complexity. We call those organs convergent, and the process that generates them convergent evolution. Canalisation is Bergson's image for it.

I think that approaching Bergson's discussion of the evolution of vision from this perspective is the best way to understand what he means by the idea that life is defined by a tendency towards indetermination, as well as how to understand the difference between developmental-organisational simplicity and material-structural complexity. In addition to philosophers of biology working on the structure/function relation, especially in connection with the convergent evolution of complex structures, this section should be especially interesting for historians of embryology, as canalisation plays an important part in the modern formation of that science through the work of C. H. Waddington.

Finally, convergent evolution provides Bergson with what he takes to be empirical evidence for his philosophy of evolution, what I consider to be the idea of unified virtual tendency in particular. After filling out Bergson's explanation for convergence on the basis of canalising tendencies dissociated from an interpenetrating whole, I situate his position in connection with some contemporary research on the convergence of complex eyes across mammals and cephalopods. I conclude by comparing Bergson's position with the science of deep homology, conservation, and developmental constraint, and demonstrating the affinities between them. The implications of this final comparison can be summarised in the following three claims: (1) adaptation is not the sole or only agent at work in evolution; in fact, it is better understood as a negative filter or contouring force operating on a more primary production; (2) selection does not act upon a wholly random proliferation of variations, but rather on the directionality imparted to evolution by the tendencies that trend it and the developmental history that it retains (through conservation and developmental constraint); and, finally, (3) there is a deep unity underlying the extant plurality of life forms, which reveals itself in the convergent evolution of like characters across divergent evolutionary lineages. Each of these three claims remains challenging, and arguably viable, for the contemporary study of variation, homology, and convergence. Bergson's 'psychological interpretation' of tendencies, their virtual modal-mereological position, and their implication within a theory of evolution for which the reality of time is the defining moment and proper object, are what build these claims out into a full-blown philosophy of biology adequate to the biology of today.

1

The Actual: Mechanism, Finalism, Modality

There are a number of ways to begin a text on Bergson's philosophy. The typical tactic is to introduce his theory of time and distinguish it from conceptions of time popular across the history of philosophy and science. Bergson begins his career by uncovering the nature of time, what he calls duration, in lived experience. His philosophy of biology deploys the idea of duration both critically and constructively. Critically, Bergson argues that it is the durational nature of life that the biological sciences deny or obscure. Constructively, it is the durational nature of life that Bergson's own philosophy of biology aims to capture. I have decided to begin by mapping out the critical elements of his philosophy of biology. That way it is clear what it is that other accounts lack, and what it is that motivates Bergson's own positive philosophy of time. Later, in Chapter 4, I show how the theory of duration as lived time can be interpreted as a component of Bergson's philosophy of biology rather than as its starting point.

Beginning the book by identifying and explaining the critical moments of Bergson's philosophy of biology allows me to introduce some metaphysical terminology I use to develop my interpretation of Bergson's own metaphysics. I represent Bergson's criticisms as arguments against the metaphysics of 'actualism' and 'possibilism' in mechanist and finalist biology. Actuality and possibility are modal categories, types of existence. Mechanism relies on a metaphysics of actualism to the extent that it regards determinate organic parts as exhaustive of the phenomena of life. Adaptationism is evolutionary mechanism. Finalism relies on a metaphysics of possibilism to the extent that it assumes that developmental ends or evolutionary events can be given at the outset of the processes that lead to them. The chapter concludes by tracing Bergson's criticisms of these paradigms through some contemporary iterations of actualism and possibilism in gene-based understandings of development and evolution.

Introduction

Much of the argumentation of Bergson's *Creative Evolution* is critical. The targets of Bergson's criticisms vary between adaptationist or mechanistic modes of evolutionary explanation and finalistic or teleological accounts of evolution and development. The objects of those accounts include the organisation of individual organisms, the source of variation and novelty in the evolutionary process, and the convergent evolution of complex structures such as the eye in cephalopods and vertebrates. Bergson does not often present his criticisms in linear argumentative form or independently of their particular objects. It is occasionally unclear whether the criticisms are meant to stand on their own account or have to be understood as parts of Bergson's positive philosophy. Neither is it clear whether Bergson's arguments can be thought to track the progress made by their targets (adaptationism, for instance) after him. It is usually easy enough to see what the targets of the criticisms share in common: a denial of the positive or effective role played by time in evolution and development. I use the term 'actualism' for conceptions of evolution that deny the effective reality of time and assume, in Bergson's words, that 'all is given' (CE 45).

Mechanism and finalism are two variant forms of the same actualism (see François 2010: 55–61). By denying the reality of time, actualism assumes that being is identical with actuality, or manifest givenness in space and time. Biological actualism is the position that spatiotemporally determinate organic parts are exhaustive of the phenomena of life. Evolutionary actualism consists in the addition that what looks like the emergence of novelty over time can be derived from the current state of things, like clockwork. Mechanism negates the reality of time by isolating the organism's parts and processes in the present and placing them in a passive causal relation with selection pressures from the outside, on the model of an unorganised material system. Finalism negates the reality of time by determining the future in advance, in the form of a developmental plan, genetic program, blueprint, or functional end, all of which are given at the outset in the mode of the possible.

This chapter reorganises the critical facets of Bergson's philosophy of evolution into a sustained argument against actualism in each form. Then it updates that argument according to the progress made by gene theory since the early twentieth century. Section 1 focuses on Bergson's critique of the mechanistic paradigm and its implication first in adaptationism and then in the more recent developmentalism that has come to complement it. Section 2 focuses on Bergson's critique of finalism. I find in Bergson's contention that 'finalism is external or it is nothing at all' a rebuke against the inner purposiveness prized by the German Idealist tradition. Then I reconstruct an argument against inner purposiveness on Bergson's behalf before turning to his more familiar critique of the theory of external finality. The last section of the chapter tracks Bergson's criticisms of actualism and possibilism through the genetic determinism that would succeed *Creative Evolution* and play a role in antiquating it for popular scientific opinion.

Mechanism

Philosopher of science John Dupré defines mechanistic explanations as explanations that 'identify the various constituent things that interact to generate the phenomenon' in question (2017). 'Arrangements of constituents with particular functions', on this view, are what 'constitute *mechanisms*'. The position that living things are mechanically explicable 'sees living systems as composed of things arranged in a hierarchy of mechanisms'. Though this definition does not pretend to be able to accommodate the various alternative accounts of mechanism on offer now (see Levy 2013 and Machamer et al. 2000), it does a good job of capturing the conception that Bergson had in mind.

A mechanistic explanation of organic systems does not appeal to anything outside of the the causal relations obtaining between the parts of the system and their local environment (see Menzies 2012 and Nicholson 2012). Mechanistic explanations do not need to attend to the history of changes undergone by the system in question, nor do they need to invoke possible 'ends' or 'aims' in view of which the system's patterns are to be conceptualised. A mechanical system – a system whose structure and operation is captured by a mechanistic explanation – is blind and non-purposive, productive of the phenomena in question without 'intending' to be. Mechanistic explanations are also presentist in that the ingredients relevant to their explanatory success are given, local, and actual. Explanatory mechanisms are finally system-independent. To identify a mechanism operative in one system is to isolate some set of components or rules of interaction that can in principle apply elsewhere as well.

Bergson understands mechanism as the product of a certain evolutionary tendency that has calcified in human intelligence. Intelligence is a theme that I return to later. Its defining tendency is the tendency towards geometrical cognition, which represents its objects spatially and in analytically decomposable form. The tendency towards this form of intelligence corresponds to and extrapolates from a tendency inherent in matter. On Bergson's view – which I explain in Chapter 4 – matter is composed of 'elementary vibrations, the shortest of which are of very slight duration, almost vanishing, but not nothing' (MM 201). The pasts and futures of these vibrations are immediate enough to be approximately subsumed within their present iterations for pragmatic purposes. It is on that basis that they can be fixed, modelled, and calculated.

The condition for the operative success of intelligence is the abstract suspension of time and the reduction of matter to sets of repetitive instants localised on a spatial grid. Intelligence thus works by deploying a set of heuristics derived from the spatialising tendencies of unorganised material systems. Mechanism proceeds on that model. Bergson's argument is that intelligence is overextended in the subsumption of life under the mechanistic purview. This is because time plays a preeminent role in the patterns and processes that typify organic bodies. Mechanism is supposed to be inadequate to the study of evolution as a consequence. I say inadequate, or perhaps inappropriate, but not false, as Bergson would have endorsed J. H. Woodger's contention that 'it is always possible to defend microscopic mechanism in principle . . . by making your mechanism complicated enough, and by "postulating" enough sub-mechanisms to meet all contingencies. It *cannot* then be refuted, but neither can it be verified' (1929: 485).

Bergson was writing in a scientific context for which evolutionary mechanism meant Neo-Darwinism. Neo-Darwinism holds that the organisation and evolution of life can be explained primarily through natural selection. Neo-Darwinism is mechanistic because it conceives the organism on the model of an inert system in mechanical equilibrium whose parts can be analytically decomposed and 'selected' in isolation from each other by environmental pressure. Since this negative account of evolution by selective elimination comprises its primary explanatory principle, Neo-Darwinism wants – as Bergson argues – for a notion of a positive account of variation and the development of complex structures. This is crystallised in Bergson's insistence that Neo-Darwinism bases its study of life on a mechanistic image of nature that it derives from the behaviour of unorganised systems. Present and future states are on its account necessary consequences of past states and 'all is given in advance', which means that a 'true evolution' can never take place (CE 37, 45).

a. Adaptationism

Peter Godfrey-Smith distinguishes between three kinds of adaptationism: empirical, explanatory, and methodological (1999). Empirical adaptationism is the view that privileges selection as the most important causal force in accounting for the outcomes of evolutionary processes. Explanatory adaptationism privileges the sorts of questions that selection-based explanations are able to resolve, such as how organisms came to exemplify design. Methodological adaptationism is the pragmatic recommendation that biologists look for features of adaptation

when they approach biological systems. Bergson's view of adaptationism tracks the first and second categories. They entail the third, though
one could accept the third without commitment to either of the first
two. When I use the term, I refer – as Bergson would have – to empirical adaptationism primarily.

The view that natural selection comprises the primary dynamic
of biological evolution is more commonly known as panadaptationism (see Barros 2008; Skipper and Millstein 2005; Havstad 2011).
Panadaptationism is the position that all biological traits can be explained
in terms of the fitness benefits they confer as adaptive responses to
environmental pressures. Bergson argues that this view is weak, retrospective, ad hoc, and improbable. It is weak because it attempts
to explain all of evolution in terms of a single mechanism, selection,
when in fact there are a number of other dynamics at play. It is retrospective because environmental pressures do not pre-exist the organisms whose traits they are supposed to explain. It is ad hoc because the
reason why a trait evolved can be construed in any number of ways
that make it look like an adaptive advantage. Finally, it is improbable.
It is improbable for any complex structure to have been constructed
out of an accumulation of random variations filtered according to utility alone. It is improbable for such a structure to continue to evolve
in complexity across a long series of generations. And it is improbable
for distinct lineages to evolve the same complex structures that realise
the same functions when they do not share either recent ancestors or
environmental pressures in common.

Each of these charges is implicit in Bergson's idea that selection is
a negative operation that explains only part of the nature of evolution. Selection works by filtering out harmful variations and preserving advantageous adaptations as a consequence. Harmful variations are
selected against because the organisms that they characterise are supposed to have a harder time surviving relative to others and to environmental pressures, which means they are less likely to reproduce and pass
along the detrimental traits. Only the traits that confer some advantage
are preserved across subsequent generations. Over enough time, after
many small but beneficial modifications are accumulated, larger scale
transformations in complex structures can occur. So long as variation is
always happening, and so long as advantageous variants are preserved,
no positive mechanism is required to explain the evolution of biological
characteristics.

Bergson doubts the explanatory sufficiency of the negative model. His
view is that selection pressures act as contouring agents, setting parameters, providing necessary conditions and constraining what is feasible

and advantageous. But they are not causally primary in the evolution of the forms adapted to them. Selection is not the engine of evolution, but its filter. Bergson analogises selection to the topology of a hilly landscape (CE 102). Any road constructed through it will have to conform to the landscape, but that conformation does not explain the form of the road, which is a route to some destination first and foremost. Evolution has its analogues to this sort of aim-orientation as well. They are developmental constraints, a default for conservation, the exaptive repurposing of existing structures for new ends, and the tendency towards increasing action. Since adaptationism leaves the positive dynamics out of the account, its explanations for how and why biological structures evolve capture only part of the nature of life. They get something right about how evolution results in particular traits, but they miss what is productive, directional, and creative about the process. I return to the positive ideas in the last two chapters of the book.

Besides being partial, the explanation of evolution by adaptive benefit alone is retrospective as well. Forms are supposed to arise as 'solutions' to the 'problems' posed by environments (see Darwin 2003: 80). Environments, or niches, are defined by what is relevant to the organisms in question. The trouble is that what is relevant is picked out by the phenotypic traits of those organisms in turn. These are precisely the traits that are assumed to be adaptive solutions to environmental problems. The selection pressures that appear to explain why some form evolved and what benefit it confers are identified on the basis of the form itself. As Niche Construction Theory establishes, organisms and environments are in dialectical relation and cannot be understood in abstraction from the way the one transforms the other. Organic forms do not mirror environmental pressures but respond creatively, crafting their own niches as a result. I return to this idea in the section on developmental constraint below, and I explain the charge that adaptive explanations are ad hoc in a discussion of Theodore Eimer and directed evolution in Chapter 5.

Bergson has the most to say regarding the improbability of adaptive explanations for the construction, evolution, and convergence of complex structures such as the eye. In a popular passage, Darwin concedes the improbability:

To suppose that the eye, with all its inimitable contrivances for adjusting the focus to different distances, for admitting different amounts of light, and for the correction of spherical and chromatic aberration, could have been formed by natural selection, seems . . . absurd in the highest degree. (2003: 172)

Darwin resolves the problem by describing what an adaptive explanation would require, then demonstrating that its requirements already obtain. The adaptive explanation requires a series of gradations from the eye in question down to a simple and imperfect optic nerve. It requires that each gradation was useful to some organism, so that it would have been selected for its fitness benefits. The eye would have had to vary slightly in each generation of organisms, and each modification to the organ would have had to make a difference to survival. The complex eye could have been constructed one step at a time as a result.

The fossil record did not allow Darwin to trace eye development back through the ancestors of the organisms that currently exist. For this reason, he looked instead to the spectrum of extant eyes, from the camera eye in vertebrates to less sophisticated visual organs in the invertebrates. The wide variation in present complexity provides an analogue for gradual evolution. This meets the first condition. The fact that each gradation currently exists seems to satisfy the second condition as well, since harmful variations would have already been selected away, leaving only organs that are useful to the organisms that possess them. The final two conditions – that all traits vary and that the variations make a difference to survival – are established facts of evolutionary science and Darwin accepts them on analogy with artificial selection. He concludes that there is

> no very great difficulty (not more than in the case of many other structures) in believing that natural selection has converted the simple apparatus of an optic nerve merely coated with pigment and invested with transparent membrane, into an optical instrument as perfect as is possessed by any member of the great Articulate class. (2003: 174)

Variation, heritability, and the utility of each gradation in the visual apparatus suffices to explain the evolution of the complex eye over geological time on the basis of natural selection alone.

Bergson does not say that adaptation cannot explain the construction of complex organs. He says that the adaptive explanation is improbable. It leaves too much out. There are more complete and less unlikely ways to account for the facts. The first reason why concerns the fact that complex organs are not only integrated structurally, but functionally as well. Chance variations in the component parts of a complex organ, since they are not coordinated by anything in advance, seem likely to sever the relation between organ and function, not improve it. It is difficult to imagine that a change made to almost anything in the complicated network underwriting vision could render it more and not less effective.

Though the camera eye is a sophisticated structure, it is not perfect or perfectly integrated. Hermann von Helmholtz famously wrote that if an optician were to sell him an instrument with the same sort of defects, he would complain and give it back (1873: 218–20). One of the most obvious flaws is in the contradictory optical and retinal designs. The eye is a wide-angle lens that appears designed for consistent optical quality across the visual field. But the retina concentrates visual detail in a small central area, the fovea, losing resolution towards the periphery. Optical and retinal resolution appear to be mismatched. The wide visual field, which is common to most species with eyes, seems important for basic survival-oriented tasks such as defence and locomotion, whereas the fovea appears to be tuned to predatory action, such as remote detection and fine-grained object recognition. Despite the apparent incongruities, clinical evidence shows that the optical system is difficult to improve via surgical modification.

Bergson might say that even though its 'design' appears defective in certain respects, the visual organ is still closer to a functionally opti-mised, integrated whole than a roughly coordinated set of individually assembled mechanisms. Even in its less sophisticated versions, the eye is made up of an elaborate set of networks and processes that are organised for functionality. For chance modifications to the structure of the organ to result in an improvement of its function it seems necessary to posit 'a mysterious principle . . . whose duty it is to watch over the interest of the function' (CE 67). Bergson did not know about the gene, but it is a postulate of that sort that he has in mind. He calls it the principle of biological teleology. It is as if there is a program that dictates which parts are necessary and how to coordinate them in view of the realisation of complex vision in a complex structure. Selection does not provide this form of developmental orientation and the coordination of parts according to a plan. For that something else is required, as no amount of selection against uncoordinated changes can explain the production of sudden, favourable transformations to a structure as tightly coordi-nated as the eye. Adaptationism, if it is to avoid the sort of improbability that Bergson discerns in it, seems to require an illicit teleological prin-ciple in order to explain structural modifications that issue in functional improvements in the evolution of complex traits.

The second reason why the adaptive account appears improbable concerns the view that complex organs are gradually assembled over geological time via the unguided accumulation of minor modifications. Darwin conceived natural selection as a 'power always intently watching each slight accidental alteration in the [elements of the eye]; and carefully selecting each alteration which, under varied circumstances, may in any way, or in any degree, tend to produce a distincter image' (2003: 175).

He imagined that each new state of the eye would undergo millions of slight variations across a long series of generations and that each would be preserved until supplanted by a superior version. Eventually, without any direction, a camera eye would result. There are two things improbable about this account. The first is the supposition that every modification to the eye that improves its function is automatically retained. It is not obvious that slight improvements in the visual apparatus of an organ would confer an advantage over competing organisms sufficient to preserve the change in every case. The improbability increases when we consider that all biological characters vary in this way. Organisms whose eyes are slightly better than others may also have developed slightly softer bones, or slightly worse olfactory senses, or slightly slower runs, or any number of other traits that would prevent them from passing on the improvement to vision. Eyes do not vary while everything else remains stable, allowing the modifications to vision to be selected on their own account.

The second is the idea that the complex eye could be constructed out of millions of slight modifications made to previously existing structures, each of which was adaptively beneficial. Since adaptationism attributes to selection pressures 'a merely negative influence', Bergson argues that 'it has great difficulty in accounting for the progressive and, so to say, rectilinear development of complex apparatus' (CE 56). Every instance of the eye seems like a functionally integrated whole of its own, not a slightly modified version of a prior model. Each improvement seems like a positive step forward, not the side-effect of the elimination of defects. The idea that a photoreceptive cell was modified over time in order to bring about a camera eye is one thing. The idea that every modification made for an adaptive organ of its own is another. It is as if we are to imagine constructing a cathedral from out of a nest by modifying millions of nests in millions of ways, arriving at the cathedral through millions of intermediaries, each of which out-competes others by improving upon them, without having the goal of a better structure in mind in advance.

The third reason why Bergson considers adaptationism improbable follows from the second, but involves a mistake. Bergson attributes to Darwin the idea that initial modifications are 'insensible', making no difference to the organ in question until a sufficient number of changes have been accumulated. In every currently existing instance of vision, the eye seems to be roughly optimised as it is, or at least dependent on a lot of parts working together. It seems like everything would have to change in concord lest isolated variations detract from the functional integration of the whole. For this reason, Bergson supposed that initially minor modifications might be so slight as to be inconsequential – making no difference relevant to the operation of natural selection (CE 63). The idea is that differences that make a difference might emerge out of the slow accumulation and preservation of those that initially do not. They could wait, as it were, for the gradual amassment of complementary variations before beginning to exert a transformative force on the structure in question. Then the organ would be capable of raising its function to a higher degree of sophistication.

Darwin himself did not claim that variations were insensible in this way. Variations could be extremely slight, but if they were strictly insensible then natural selection could not act on them (Darwin 2003: 80). It is possible that Bergson misread Darwin on this point, misunderstanding the claim that natural selection is 'silently and insensibly working, whenever and wherever opportunity offers, at the improvement of each organic being in relation to its organic and inorganic conditions of life' (2003: 82). The action of natural selection is what is insensible because it is negative, merely filtering out what is harmful and so ultimately invisible in what is preserved. But variations, if they are to provide material for selection, have to make a difference. Bergson may also have confused Darwin's species gradualism, according to which the differences between forms blend into each other along an 'insensible series', with a position on the individual differences themselves (2003: 54, 158). It is also possible that Bergson inherited the idea of insensible variation from the readers of Darwin that he read (see Tahar-Malaussena 2020). I suspect, however, that Bergson was drawing what he understood to be the requisite conclusion on Darwin's behalf. Either variations are slight but always sensible and more likely to disrupt the functional integration of complex organs than improve it, or else the variations are initially insensible and can accumulate in the direction of large-scale transformation without risking disruptive effects at the outset. Bergson likely considered the latter to be the more reasonable view. He seems to have found the former difficult to reconcile with the existence of integrated structures at all, if they are always randomly varying and if every variation made a relevant difference to them.

The problem with the idea that variations are initially insensible is that natural selection would be unable to explain how they could be retained. It cannot be the case that the initial variations are preserved with a view towards the final product, as if the species knew in advance that it needed to improve its power to see and so began selecting for initially useless variations in its visual apparatus that, when taken together at a later date, might eventually prove effective. That is the kind of teleology that adaptationism has to circumvent. Since the

organ has to be understood in terms of its history of development – as
I discuss below – the evolution of vision would require that the same
complex set of initially insignificant variations occur and are preserved
in more or less the same order along every evolutionary trajectory that
results in complex eyes. If adaptationism is understood in this way,
then it contradicts itself. Variations cannot be accumulated in view of
an improvement 'when each of them, taken separately, was of no use'
(CE 65). Utility is natural selection's only operative standard (Darwin
2003: 77).

Some adaptationists reject Darwin's gradualism and defend the
position that evolution works on large-scale structural transformations
instead (CE 65). While Bergson considered this an unlikely alternative,
it was – sixty-five years later – to be reformulated as a theory of speciation by Gould and Eldridge and soon afterward accepted as the best way
to explain the extant evidence of a discontinuous fossil record. Their
theory, 'punctuated equilibrium', has two central tenets: first, that most
new species do not come to be by slowly accumulating differences in
a direction away from their ancestral populations, 'but by the splitting
(branching) of a lineage into two populations'; and second, that 'most
peripherally isolated populations are relatively small and undergo their
characteristic changes at a rate that translates into geological time as an
instant' (Gould 2001: 346, 347). Punctuated equilibrium explains the
fact that the fossil record shows abruptly generated forms followed by
long periods of stability by locating speciation in population divisions,
and in hypothesising a faster rate of change in smaller populations (see
Gould and Eldridge 2004). Species arise quickly out of isolated groups
(the punctuation) and then persist as large and stable populations, changing little (the equilibrium), until dividing, changing, and stabilising all
over again.

As an explanation for the evolution of vision, this would mean that
'the eye of the mollusk and that of the vertebrate have both been raised
to their present form by a relatively small number of sudden leaps'
(CE 65). Bergson considers this an improvement over the idea that the
eye is a product of 'an incalculable number of infinitesimal resemblances
acquired successively' (CE 65). It may be easier to see how selection
could work on these sudden leaps, since the leaps would make a significant difference to the utility of the organ in question. But it is difficult to
imagine a proliferation of functionally integrated structures at random,
only some of which are beneficial and retained. Punctuated equilibrium
may provide an attractive account of speciation, but it seems deficient
as an explanation for the evolution of functional structures. The adaptationist seems to be committed to the view that favourable large-scale

transformations are conserved through the elimination of functionally deficient variations in a sorting process that runs until a successfully coordinated organ occurs by accident and can be retained for fitness benefit. Birds have wings and eyes now, but we are presumably to imagine similar species dying off by trying to navigate their environments with sets of other functionally operational but less beneficial organs instead, in a kind of biological vaudeville.

When it comes to the development of a single organ over generations, the component parts of each increasingly complex structure still have to be produced, or rather 'hit upon' accidentally, all of a sudden, in coordination. It seems unlikely that different evolutionary trajectories would hit upon and pass through the same series of sudden improvements, in the same order, to progressively construct and converge upon an organ as complex as the eye. What appears likely or not, and why, is of course not a decisive argument for or against anything. What Bergson is attempting to show is that the prevailing explanations appear improbable because they lack a positive account of variation and a directional account of development.

Bergson thinks that a gradated series of increasingly sophisticated visual organs is better explained by the realisation of a directional potential or tendency, orienting variations in accordance with the function of vision. He thinks the existence of directional tendency is clearest in cases of convergence. Adaptationism is supposed to be at its most improbable as an explanation for these cases. Evolutionary convergence refers to the appearance of closely related anatomic parts with like functions across unrelated taxa, in different environments, by means of independent evolutionary lineages. Examples range from wings capable of flight in bats, birds, pterosaurs, and insects, to photoreceptive lenses in a wide assortment of divergent species. 'What likelihood is there', Bergson asks, 'that, by two entirely different series of accidents being added together, two entirely different evolutions will arrive at similar results?' (CE 54). The further the divergence between these two lines, the less likely the adaptive explanation for similar results. The idea is that the evolution of the eye is improbable enough in its own right, but even more so the more often it occurs in different species and different environments.

One way to make the adaptive account seem less unlikely is to suspend the causal relationship between the production of a structure and the linear history of the variations that constituted it. This suggestion resists the intuitive assumption that like structures would have to have arisen from out of like causal orders. Even if it is the case that such structures, striated along different evolutionary lineages, had to arise out of some set of causes, those causes might have differed from each other while each leading to the convergent structure in question. There may be a number of ways to build an eye, in other words, each of them nevertheless drawing on nothing but chance modifications in sequence. That distinct sequences eventuate in a convergence need not require the assumption of a telic orientation from the outset if you can get to the same place – as Bergson metaphorises the claim – by taking more than just one road.

Bergson thinks that the distinction of destination from route makes for a misleading metaphor. Realised structures cannot be abstracted from the evolutionary histories that generated them. 'An organic structure is the accumulation even of those small differences which evolution had to pass through [*traverser*] in order to reach it' (CE 56 tm). The destination – the organic structure – is its very route retained; the form of the one has to be understood in terms of the history of the other. This idea should not commit Bergson to a rejection of exaptation or spandrels (see Gould and Lewontin 2006). Even if traits evolved under one set of selection pressures can be functionally repurposed and prove adaptive under another set, it remains the case that organic structures retain their histories in the form of such things as structural and developmental constraints (as I discuss in Chapter 6).

There is a profound relationship that obtains, in any event, between the products of evolution and the accumulation and consolidation of the processes of change that condition their realisation (see Mengel 2009). Evolution 'implies a continual recording [*enregistrement*] of duration, a persistence of the past in the present, and so an appearance, at least, of organic memory' (CE 19).[1] This organic memory outstrips any one of its local actualisations in a present stage of organic development. In order to specify it fully, 'the *entire* past of the organism, its heredity, and finally the whole of a very long history must be added to that moment' (CE 20 tm). It follows, for Bergson, that the evolution of the same structure across diverse phyla cannot be explained in abstraction from the developmental trajectory of each – the pasts that they record, retain, and embody.

Richard Lewontin's more recent work on the subject recapitulates Bergson's arguments against the implicit mechanistic equivalence between organism and machine. Lewontin specifies a number of ways in which organic individuals are inconceivable apart from their contingent histories of development. The species to which they belong are similarly inconceivable apart from the 'wandering pathways' of the evolutionary trajectories through which they came to assume their current forms (2000: 88). Lewontin agrees with Bergson (without knowing it) in taking this dual fact – the contingency of developmental history and its cumulative retention in the organism – to ratify the disanalogy between living things and unorganised material systems. While mechanisms are in principle analytically separable from their causal pasts, 'it is impossible to understand the situation of living organisms without taking into account their history' (2000: 88). Understanding organs and functions requires understanding the morphogenetic processes that produced them. Understanding organisms and species requires an evolutionary account of their geneses. Destinations are, again – at least in biology – the effects and embodiments of the routes taken to arrive at them.

Genetic accounts are, however, now little more than *genetic* accounts. The genesis of the realisation of a particular trait is now traced and explained through the isolation and identification first of the genetic blueprint that regulates and controls for its development in the individual, and second of the conservation of the relevant genes across phylogenetic history, where those genes are derived in different lineages from the ancestors they have in common. Today accounts of convergence by adaptation follow suit, enlisting in their explanations the genes that are shared across the lineages in question (even if they are acquired independently), and the RNA splicing processes that leverage the information encoded in them towards developments in the structure of convergent organs.

Though Bergson was writing right around the time that Mendel's rules for inheritance were rediscovered by de Vries, Correns, and von Seysenegg, he could not have foreseen how the gene would reshape debates around convergence. There is nonetheless a point to be made in invoking this shift in the theoretical resources of the evolutionary scientist. As it will happen, the identification of the genetic bases for the evolution of convergent organs reveals a shared set of internal constraints that function to control for their development in addition to the adaptive pressures that otherwise orchestrate their evolution from the outside (see Gould 2002: 1068–89). In other words, gene theory produces – at least according to one popular subset of its theorists – a reproach to the adaptationist explanation for convergence, and so something like an authorisation of Bergson's own critical remarks as well.

b. Developmental constraint

The photoreceptive eye, first offered forty years ago as a case study in convergent evolution, remains a textbook example of convergence today. It is now discussed almost entirely by way of the conservation across divergent phyla of a single 'master control gene', the *Pax-6* (see Gehring 1996, Walther and Gruss 1991, and Halder et al. 1995). The *Pax* gene family performs critical functions in the embryogenesis of tissues and organs. The *Pax* genes provide instructions for the production of proteins, which attach to specific strands of DNA and control for the expression of other genes as well. They are also understood as transcription factors that preside over the activation of the genes associated with the formation of eyes. *Pax-6* was first cloned in mice using *Drosophila* probes in the early 1990s, and was soon afterward linked to mutations in the form and function of eyes. Further studies established that it was the expression of a homolog to *Pax-6* that induced eye development and deformation in the flies as well. Conservation between the mammalian and insect sequences turned out to be over 90 per cent. Their functional homology was then affirmed by inducing the mouse gene in *Drosophila* and finding that the mammalian version could still induce the formation of normal eyes in the fly. A genetic homology between the two phyla was uncovered in the form of a shared developmental constraint on the formation of their eyes.

The paradigm for convergent evolution has long been defined by the remarkable similarity in the form and function of the single-lensed eyes of vertebrates and cephalopods. Eyes work more or less the same way in squids as in birds. Yet the vertebrate eye is formed out of brain tissue while the cephalopod eye is formed out of the epidermis. The vertebrate eye seems to have evolved backwards and upside down, requiring light to travel first through the cornea, lens, blood vessels, and several layers of cells before reaching the light-sensitive rods and cones that transduce the signal into neural impulses that are processed into patterns in the visual cortex (see Shermer 2006: 17). This design has its drawbacks – the mammalian blind spot, most famously – but it does allow for the outer retina of vertebrates to sustain higher metabolic activities. The cephalopod eye, by contrast, seems to have been constructed the 'right way out', with the nerves attached to the rear of the retina, which means that it does not have a blind spot. A genetic account for this difference seems plausibly derivable from the developmental differences in the embryogenesis of vertebrate and cephalopod eyes. The former develops its eyes as extensions of the brain while the latter invaginates the epidermic surface area of the head.

Vertebrates and cephalopods appear to converge on the same structure without evolving it out of any of the same processes or parts, and 500 million years after diverging from their last common ancestor. Genomic sequencing has, however, revealed that an ancestral *Pax-6* likely controlled for the formation of a simple assembly of light-sensitive cells in the Cambrian period (Serb and Eernisse 2008). Over time, the 'instructions' that arose from differently spliced variants of a single *Pax-6* diversified and integrated a set of associated genes responsible for the formation and patterning of the membranes, muscles, proteins, chemicals, and glands that come together in the realisation of the complex eye.

Bergson maintains that 'adaptation explains the sinuosities of the movement of evolution, but not its general directions, still less the movement itself' (CE 102). Adaptation may be able to explain the adaptive differentials between advantage and drawback that are reflected in the different evolutionary 'choices' made in vertebrate and cephalopod eyes, but it cannot on that basis account for the parallel evolution of the registration of visual information in the camera-like apparatus itself.

Bergson allows that an original light-sensitivity of the kind identified in the Cambrian period may have arisen out of a response to the problem of how to navigate a light-saturated environment. That problem may have selected for the evolution of cells receptive to light. The initial cell-assembly can even be conceived as an 'imprint' of its environment, a mould conformed to it – which is the position assumed by the adaptationist explanation for convergence. But once those cells are functionally integrated with an organic body that prolongs the perception of light into movement, enlisting a series of behaviour programmes, a basic somatic memory, and a variety of organic systems in its sensorimotor activity – and once that larger functional system starts to evolve in complexity in parallel across divergent phyla – then it can no longer suffice to speak of light as an environmental problem that drew its own solution into realisation. The complexity of the solution outstrips the problem that prompted it, marking a difference in kind between them.

Bergson locates a bifurcation moment in the apparently linear progress from the original pigment spot to the camera eye. There is a difference between the initial 'insinuation' of the tendency towards vision in the pigment spot – which does conform itself to external pressure in such a way that it makes sense to speak of a passive adaptation of organ to environment – and the subsequent development of the pigment spot through a series of intermediaries towards the complex eye. This distinction is one of a difference in direction, and so represents, on Bergson's system, a difference in kind as well. The organ-to-be originates in a material patterning, in much the same way that an orator begins by falling in 'with the passions of his audience', and at this point photoreception as a passive response to the influence of light is indistinguishable from the tendency towards the complex exploitation of light for locomotion (CE 71). Once properly insinuated there, as when the orator has made himself coincident with the passions of his audience, the tendency towards vision is then able to leverage its original patterning in the direction of the eye, as when the orator begins to turn the audience's passions towards his own purpose, passing from *following* to *leading* them. What looks like a linear line of development covers over a change in direction. The evolution of the complex eye requires that directional shift, from the initial passivity of adaptation to the progressive realisation of an aim-oriented activity. I return in Chapter 5 to the idea that vision is a functional end constraining the evolution of visual organs.

Bergson finds this directional difference reflected – albeit equivocally – in the adaptive account, in the difference between passivity and activity. In the case of the eye, it is a question of where to locate the evolutionary agency. Adaptationism begins by locating it in selection pressure, conceiving the eye's development in passive terms. The adaptation of an organism to its environment is like the conformation of a liquid to its container. The form is there in the container (the environment), and it reflects itself in the content upon which it is to be imposed (the organism). This explanation seems to account for initial developments, like the first appearance of a set of photoreceptive cells in the lower Cambrian period, about 540 million years ago. But from there any increase in functional sophistication has to enlist another explanatory step, for while there is in the initially photoreceptive cell something like a passive reaction to the influence of light, no such mechanical correspondence holds for the cell's increasingly intricate descendants.

The first point should not be conceded too quickly either. It should be asked whether an analogue to the ancestral photoreceptive cell has ever been successfully generated in a model organism as a response to a light-saturated experimental environment. The functional unit of photoreception involves opsin proteins that initiate a nerve impulse in response to light along a hairy layer derived from extensions of the cellular membrane in protostomes, and from the separate structures of the cilia in deuterostomes. The question is whether this basic system has ever been experimentally generated independently of the already existent genetic machinery that controls for the evolution of photoreception and the developmental pathways attendant upon it. If that machinery and those pathways are supposed to have been derived in present organisms from the ancestral functional unit itself, then the experimental confirmation of the merely passive evolution of original photoreception has to remain either speculative or circular. We cannot recreate the conditions that are supposed to have given rise to photoreception without employing in their recreation what are supposed to be the products of those very conditions.

Adaptation cannot be thought of as an affair of *repeating* environmental pressure in organic terms at any stage in the evolution of the eye later than the development of the first photoreceptive cells. At later stages, the evolution of the eye has to be conceived in terms of *replying* to those pressures, modifying them in turn (through the dynamics of niche construction), and creating in response a form that neither resembles nor is prefigured by its environment. The question, then, is how it is that complex developments in the structure of an organ, even if they began by mechanically repeating the form of the problems posed to them, are supposed to generate the same solutions over time in different phyla. Bergson locates the adaptationist equivocation here, rendered in the terminology of repeating and replying. It can be translated back into the distinction between passivity and activity. The mechanistic explanation for the convergence of complex eyes begins by modelling their origin as a passive imprint of the environmental influence of light. It proceeds from there, by identifying the genetic mechanisms that underwrite that first passive imprint, to shift from the passivity of an imprint to a set of developmental possibilities opened for the organism by the genes that preside over the embryogenesis of its photoreceptive cells. As those genes are spliced, recombined, and integrated with others, they work to explain the evolution of the eye in excess over the environmental influence to which it first conformed. Mechanistically reactive to its outside at first, the organism gradually assumes the traits of an active cause in its own evolution.

As the products of this active cause begin to trend in the direction of a sophisticated camera-like eye, not only is the originally passive sense of adaptation left to one side, but a new kind of 'intelligent activity, or at least a cause which behaves in the same way' is surreptitiously introduced into the explanation for how the eye has come to be approximated across divergent evolutionary lineages (CE 58). Adaptationism is supposed to be mechanistic, and therefore not intentional or intelligent, but it is a mechanism that can nevertheless be thought, in a Bergsonian spirit, to recuperate an equivalent 'cause which behaves in the same way' (as an intelligence), in the agency of the gene in particular. In this way the fact that more complex configurations of the eye form according to the same genetic cascade identified in the ancestral visual system is taken to indicate the pre-existence of the genetic pathways as a positive constraint on the evolution of the eye to begin with (Pineda et al. 2000: 4525).

While studies continue to proliferate, many researchers now accept not only that there is a shared basis in the developmental pathways for complex eyes no matter where the eyes arise, but that the pathways – as positive constraints or facilitating channels – work to explain the widespread convergence of eyes in divergent lineages such as vertebrates and cephalopods as well. Stephen J. Gould draws the polemical conclusion: 'evolutionists can no longer argue that such similar eyes originated along entirely separate routes, directed only by natural selection, and without benefit of any common channel of shared developmental architecture' (2002: 1128). Bergson could have written those same words. He even employs the phrase 'parallel development' in the place of 'convergent evolution', the same terminological move made by Gould to signal the internally directed evolution of the same organ in *parallel* across different lineages in contradistinction to the heterogeneous evolutionary *convergence* that conceives like solutions to be separable, in principle, from a shared set of developmental pathways and available in their entirety to external pressures.

The difference between parallelism and convergence consists in the differences in context between explanatory causes (see Sansom 2013). Constraints refer to causes imported from outside the local context to explain observed patterns by drawing on how they might be controlled and defined by their developmental histories. Selection pressures are local to the patterns observed, operating on the population in question to furnish the traits at hand. Convergent traits are homologous if their parallel development is internally regulated by shared constraints; they are analogous if they are selected for as solutions to the same environmental problems, converging from out of independent histories. The point is that if the evolution of the eye proceeds on the basis of a shared set of genes that regulate for its patterning and development, and if these genes have been conserved across the animal kingdom from their origin in the first set of photoreceptive cells, then it seems as if adaptationism has to be mistaken at least to the extent that there is good reason to suppose a common evolutionary origin for the complex eye in each of its appearances.

I do not think it goes too far to suggest that Bergson had in mind the outlines of this sort of position when he declared the adaptationist hypothesis vitiated by extant biological evidence. He argues that there must be something that directs the parallel development of convergent structures towards their realisation on each line. A similarity of external pressures is insufficient. The developmental view takes a welcome step in the right direction, but it goes only so far as to conceive of a *constraint* on the pathways leading to the parallel formation of convergent structures, and so nevertheless lacks a positive principle that would explain the complexification of the pathways and their eventuation in the same complex organs. Developmental constraints are, moreover, inherited genetically, which is to say that they remain attendant upon the same biological actualism that underwrites adaptationism, the position that everything relevant for the explanation of its evolution is given in the actual organism (or at least in its genome). The constraint may indicate a deep homology running beneath divergent phyla and resurfacing in examples of the parallel development of like organs, but it is nonetheless given in the genetic blueprint of each. Time, once again – at least understood as Bergson understands it, as the retention of a history of change and the necessary unforeseeability of its future – seems to have been abstracted from out of this conception of evolution and plays no positive role in the explanation of how new forms evolve.

If the genetic mechanisms identified by the developmental view are thought to control for the evolution of parallel structures in advance, then this is perhaps the point at which mechanism lapses most clearly into the finalism that it is supposed to be deployed over and against. Later in this chapter I consider some ways in which the gene theory invoked in accounts of developmental constraint might be said to operate upon some of the same suppositions that Bergson identified in the finalist biology of his era. The next section reconstructs Bergson's criticisms of finalist biology, the closed conception of the organism, and the implicit metaphysics of possibility or 'possibilism'.

Finalism

Finalism is a theory about ends or purposes. To regard an object finalistically is to predicate purposiveness of it, explaining that object through its purposiveness, goal, end, sense, or plan. Finalist theories are determined and differentiated through object-individuation: the specification of what it is that is to be understood purposively on their account. Some philosophers of biology are finalists about organs by maintaining that there are normative constraints on how we understand organs to function. Others introduce finalist concepts into talk of the natural selection for beneficial traits. Others follow Kant, ascend one level of organisation, and ascribe teleology to organisms qua self-organising systems. Others are finalists about ecosystems and larger group concepts. Before the nineteenth century it was a widespread belief that the concept of nature included finality, telos, end, purpose, meaning, plan, and sense as its requisite predicates. Since finalism in general is applicable to so wide a range of objects, in so wide a range of registers, it is a theory less rigidly fixed than mechanism; as a consequence, it is practically irrefutable, since it can always recuperate itself in a differently individuated form. That finalism is so multi-variant will prove an important point for Bergson's own evolutionary programme. To the extent that one rejects mechanism, one accepts, he argues, some form of finalist explanation by necessity.

Bergson admits two basic forms of finality – local, 'internal', or Kantian; and global, 'external', or Leibnizian – before arguing that they cannot be held independently, and that there is only one viable form of the theory to be found in the conceptual conclusion of its core premise. Bergson locates its clearest expression in the Leibnizian doctrine that 'beings merely realize a programme previously arranged' (CE 39).

Leibniz serves to personify a certain position within Bergson's critical programme (see François 2010: 46–9). That position is the elaboration to its conceptual conclusion of the idea that actual entities are the realisations of possibilities that pre-exist them. To the extent that the reality of time requires a concomitant epistemic unforeseeability, any defence of pre-existent possibilities is as mistaken as the extreme form of the position that Bergson attributes to Leibniz. On this account, the event of any actualisation was preceded not only by its own specific possibility, but by a global set of possibilities together comprising the plan or programme on the basis of which it was realised. Bergson calls this 'external' finalism (I discuss this term below), and maintains that it is the only legitimate form the theory can take. It is on the basis of that conception that he reduces finalism and mechanism to their shared rejection of the reality of time. Bergson does, however, locate in the Leibnizian theory another tenet, which consists in the equation of organisation with purposiveness. Once dissociated from the first, the second tenet was developed into a new doctrine of finalism altogether, the theory of inner purposiveness.

Though Bergson derives two main tenets from Leibniz's finalism, the theory itself has three primary components: complete concepts, compossibility, and the divine decision (see Mates 1972; cf. Jorati 2017: 59–91). Since individual substances are in principle modally indeterminate – they can be either actual or possible, necessary or contingent – they cannot be defined by any present instantiation of their properties. They are defined instead by their concepts. While the question of the modal status of an individual is in principle open, its concept has to be instantiated somewhere. For Leibniz, that means 'in a certain realm of ideas, so to speak, namely, in God himself' (1989: 151). God, because he is omnipotent, has what Leibniz calls a 'complete individual concept' of every possible finite substance, which contains every predicate that is, has been, and will be true of it. The complete concept is the essence of an individual. God is able to survey all possible individual essences and arrange them in sets defined by 'compossibility'. Compossibility refers to possible essences that can coexist together within the same logically and lawfully consistent world. Compossibilities are combinations of possibilities, and there are exactly as many possible worlds as there are sets of compossible individuals. The question is how it has come to pass that one of these possible worlds was actualised. Leibniz's answer is the divine decision, God's act of actualising the 'best' set of compossible individuals. Because of his benevolence, it is in God's nature to select the best of all possible worlds. The best possible world is for that reason the actual one.

The divine decision bestows a global purposiveness upon the actual world as a whole (see Hartz 2011). This is the second tenet of Leibniz's finalism, what Bergson calls its 'externalism', for it supposes that 'living beings are ordered with regard to each other', that is, from the outside of each (CE 41). This second tenet is based on the first, which consists in God's possession of the complete concept of each possible individual. On Bergson's gloss, every actual being realises a programme set out for it in advance. According to Leibniz, the programme is held in the timeless eternity of the divine intellect. These two tenets can, as I said, be dissociated from each other and independently developed. The first – possibilism – is retained by theories of genetic determinism, while the second – purposive organisation – ramifies away from its original globality or externality and towards what Bergson takes to be a problematically individuated internalism instead.

a. Inner purposiveness

Bergson understands the local or internal account to have arisen as a consequence of the empirical difficulties faced by the global externalism that he attributes to Leibniz. It is, as Bergson says, 'the tendency of the doctrine of finality' to 'thin out the Leibnizian finalism by breaking it into an infinite number of pieces' in response to the fact that 'if the universe as a whole is the carrying out of a plan, this cannot be demonstrated empirically' (CE 40). Indeed, 'the facts would equally well testify to the contrary', as nature taken as a whole seems to evidence as much disorder as order, chaos as harmony, as much retrogression as progress (CE 40). Though it seems unlikely that finality might be reasonably affirmed of the whole of life as such, 'might it not yet be true, says the finalist, of each organism taken separately?' (CE 40). By further individuating its object, finalism locates the empirical reality of disorder in the clash of organisms with each other in order to preserve purposiveness at the level of each individual organism taken with respect to itself. Bergson calls this 'internal finality' since it attributes purposiveness to the internal composition of the organic body as an explanation for the division of labour among its parts and their integration in service of the end of the individual whole.

When Bergson declares this 'the notion of finality which has long been classic', he has Kant and Hegel in mind (CE 41). Kant distinguished between external and internal purposiveness in the same terms that Bergson employs for his own distinction in theories of biological finalism. By external purposiveness, Kant meant the finality of artifactual manufacture, since the particular end served by the artifact lies outside of itself in its use. By internal purposiveness, he meant the particular kind of finality that qualifies living beings, for the ends served by the organisation of their parts are *internal* to the wholes that they compose (2000: §82; 5: 425). The living being is to be conceived through itself, as self-organising. It is, as a whole, the final cause of the efficient-causal relations among its parts, even as it is constituted by them recursively. Yet the thought of purposive organisation nevertheless requires the thought of an external intention. This was the central antinomy of Kant's 'Critique of the Teleological Power of Judgment' (2000: §70; 5: 387). Its resolution was to come by way of Kant's account of regulative judgments. Teleology was not to be predicated of the organism constitutively, as if it really was a made thing. Teleology was rather to be ascribed to it only regulatively, as a necessary constraint on the intelligibility of the organism as organised matter.

Kant maintained that the organism was irreducible to the physico-chemical processes that constitute it. While an organism's parts may be understood as products of mechanistic laws, 'the cause that provides the appropriate material, modifies it, forms it, and deposits it in its appropriate place must always be judged teleologically' (2000: §67, 5:378). The organism is something over and above the parts and processes that make it up. It is only the assumption of teleology that can preserve the organism itself, as a whole irreducible to its parts. It is, however, at one and the same time to be regarded naturally or non-artifactually. The ambition of the 'Critique' was to furnish a resolution to the antinomy between the obligation to think organisms as purposes and the concept of purposiveness itself, which includes the thought of an external intention. Kant's resolution was to nuance the judgment of purposiveness by distinguishing between its regulative and constitutive attributions. While constitutive judgments of purposiveness are made with respect to artifactual productions, when it comes to living things purposiveness can only be postulated regulatively. We can assume purposiveness in an organism 'even if its possibility does not necessarily presuppose the representation of an end, [. . .] because its possibility can only be explained and conceived by us insofar as we assume as its ground a causality in accordance with ends' (2000: §10, 5:220). For an organism to be a purpose just means that we require the assumption that it was produced according to design in order to understand it. We necessarily regard living things as ends not because we can infer from them the existence of an external intention, but because without assuming such an intention we are unable to understand them.

Hegel's praise for Kant's inception of the idea of natural or inner purposiveness is as prevalent in his writing as is his censure and rejection of the externalist residues of the conception of purpose that Kant maintained. Purposiveness could only be properly internal, on Hegel's account, if it was first liberated from its tie to intentionality, and then ascribed to the organism itself. This means that the organism really is cause and end of itself, not just for us, as a matter of its intelligibility (as it was for Kant), but in and through itself constitutively (1969: §204–§212). The decisive Hegelian definition of life was to consist in large part in the non-oppositional reciprocity between organisation and purpose in the individual being (1969: §216, §219). It is Hegel, then, that best represents the internalist culmination of the second tenet of Leibnizian finalism. And so, by 1907, it would have made sense for Bergson to consider 'classic' the formulation of finalism that restricted the attribution of purposiveness to the organism as self-organising individual.

Bergson's contention that 'finality is external or it is nothing at all' may have the German Idealist philosophy of nature as its foil (CE 41). The argument can be reconstructed in four steps: discernment, criticism, and two inferences. Bergson begins by discerning in the theory of internal finality its dependence on a conception of the organism as a rigorously bounded individual. If its parts are to be subordinated to the organismal whole as the final cause of their organisation, then there must be a determinate distinction between that whole and its outside. Purposiveness can only be internal with respect to a limit that would differentiate that internality from what is external to it. The theory of inner purposiveness stands or falls with the individuality of the organism.

The second step to the argument is a criticism of this conception of the organism. It is motivated by two considerations: the relative autonomy of the parts, patterns, and processes that constitute the organism; and the continuity of the germ cells through their temporary instantiation in the soma. Bergson begins by noting that each of the elements of the organismal whole 'may itself be an organism in certain cases', that 'the cells of which the tissues are made', for instance, 'have also a certain independence', and that, in sum, there is to be attributed to the organism's parts a relative autonomy from the whole (CE 41–2).[2] The same self-organisational powers that define the organism as a whole – what contemporary theorists call an 'autopoietic system' – are characteristic too of the parts and processes that constitute it, as well as the subsystems whose interactions constitute them in turn.

Chilean biologists Maturana and Varela (1980) coined the term 'autopoiesis' to refer to the self-producing, organisational activity on which Kant based his own theory of the inner purposiveness of organised bodies. They begin with the bacterial cell as the simplest instantiation of autopoietic life. As a system, the cell produces the components necessary to distinguish it from its domain by way of a continuous process of chemical reaction. The instantiation of that distinction in the membrane organises a network of components that define the cell as a distinct entity. The components of the cell are also involved in the regeneration of the boundary responsible for producing them. The self-organising activity of the cell is the capacity to reproduce a distinct unity over time that is in principle attributable to all the cells – as well as the cellular subsystems that they compose – that make up the organism as well. The unified individuality of the organism is better understood as the coordination of a set of self-organising living systems that are each at the same time conceivable as unified wholes in their own right. The inner purposiveness of the organism is in principle *external* to the inner purposiveness of each one of its parts when they are understood as self-organising systems themselves. This is just what Bergson means when he says that 'the idea of a finality that is *always* internal is therefore a self-destructive notion' (CE 41).

In dramatising this mereological independence in the example of the phagocyte, Bergson seems again to anticipate a recent development in the research, this time in the immunological displacement of self-identity. He notes the way phagocytes 'push independence [of part from whole] to the point of attacking the organism that nourishes them' (CE 42). The phagocyte is a type of cell that helps protect the organism by engulfing and digesting bacteria, pathogens, and dead or dying cells (following apoptosis), acting as a protective filter on external and internal threats to the integrity of the whole. This process – phagocytosis – is now understood to be an integral feature of the animal immune system, but it was initially relegated to the secondary status of a clean-up mechanism, activated after the organism's 'humoral' factors had already neutralised the threat and re-established homeostatic equilibrium.

It was Metchnikoff's 1891 *Lectures in the Comparative Pathology of Inflammation* that made the decisive difference. His agential reconception of phagocytosis involved destabilising the idea of the organism as a whole. Metchnikoff reconceived organismal identity as a temporary achievement of coordination across disharmonious elements and processes. Some kind of active mechanism was required, and phagocytosis – by repurposing its ancient nutritive function as a mechanism of attack by inflammation and engulfment – was the ideal candidate for that role. While this innovation solved a number of problems, it generated one that has remained more or less intractable ever since. This is the question of how the phagocyte – as well as the immune system more generally – determines what to neutralise or destroy as opposed to what to tolerate, ignore, or assimilate. A complex history finds its first footing here, now defined retrospectively by the cognitive model first proposed by Burnet in the 1940s, according to which the system in question is regulated by a self/non-self distinction that triggers an immune reaction in response to the presence of non-self where it should not be.

The last several decades of immunological theory seem to have ratified Bergson's intuitions by recovering Metchnikoff's phagocytological displacement of 'the notion of a self, transcendentally fixed and cohesive', in terms of which the immune system can defend against intrusions, in favour 'of a conception of the organism with a dynamic identity, whose boundaries and structure evolved in the face of internal and external challenges' (Tauber 2017: 3). Recent theorists have motivated this turn away from the cognitive model's self-identical organism with observations drawn from the phenomena of autoimmunity, the environmental acquisition of the elements that constitute the immune system, and the tolerance and inclusion of bacteria, parasites, grafts, and foetuses – all of which culminate in an ecological view of the organism as individual (see Pradeu and Carosella 2006: 239–40). On the ecological view, the immune system does not regulate for a mereological internalism defined against an invasive outside, as if its only operative mode were defensive, but rather acquires, integrates, and monitors a range of microbial symbionts from its environment, evolving pattern-recognition mechanisms that control for aberrations from what comes to be a contingently emergent norm.

In noting how the mechanisms required for there to be the kind of mereological closure assumed by the theory of internal finality can undermine the final cause that is supposed to order them, Bergson placed himself in line, in advance, with the challenges that immunological research was to put to the idea of a self-identical organism in the century that followed him. Bergson made the same point in terms of the impossibility for individuation, identity, or mereological closure to fully establish itself in the organic domain. This formulation is represented in the second reason motivating Bergson's argument against the internalist conception of organism. Given the facts of reproduction, organic individuation is always incomplete.

Bergson makes passing reference in the context of his critique of internalism to what theorists now call the 'Weismann barrier', the theoretically inviolable division between germinal and somatic cells (CE 42). August Weismann held that it is only germinal cells, or gametes, that have heritability functions (1891: 174). They pass information along their own line only. The somatic cells are a product of the totipotent zygote, which is a product of the fusion of haploid gametes or germ cells. The germ cells are formed on the basis of a vital substance that Weismann called the germ-plasm, which remains continuous and unchanged through each iteration of this process. The causal line runs in one direction. The germ cells give rise both to themselves as well as to the somatic cells, while the somatic cells produce only cells that develop into the body of an organism. This internal split rends the organic world in two, subtracting evolutionary significance from constituted organisms and restricting it to the pre-individual germ line that runs through them. This means that somatic mutations cannot be inherited. Neither can habits, acquired characteristics, or associations. The germ line is deathless. Individual organisms are its temporary excrescences, epiphenomenal by-products deposited along the course of the germinal flow. This is an extreme position, and has since been complicated by the epigenetic revolution in evolutionary theory. Bergson thinks, in any case, that reproduction is on its own sufficient to furnish his conclusion that individuation is always necessarily unfinished.

Bergson turns to reproduction to dispute an over-individuated form of vitalism according to which every biological individual is organised by a vital principle of its own. The adult organism 'is only the development of an ovum', and the fertilised egg, inasmuch as it is formed by 'part of the body of its mother and of a spermatozoon belonging to the body of its father', is 'a connecting link between the two progenitors' (CE 43). Bergson concludes 'that every individual organism, even that of a man, is merely a bud that has sprouted on the combined body of both its parents' (CE 43). The 'bud' is another Weismannian term of art, designating the fertilised egg as it was thought to contain the genetic 'determinants' necessary to the development of the organism (Weismann 1893: 53–60, 154–63, 184). The egg is the 'unopened bud' of an adult individual. While the 'bud' was supposed to open or 'flower' progressively through its division into the cells of the developing organism – Weismann's 'blastomeres', from the Greek term meaning 'part of a bud' – it was thought to form or 'sprout' originally from out of a portion of the parental germ-plasm that remained unchanged (and inactive) in the nucleus of the egg-cell. Fertilisation introduces another set of what Weismann called 'ids', higher order determinants, into the egg-cell, directing the realisation of the organism through the ontogenetic processes of cellular division and differentiation. This is what Bergson means when he says that the organism is a bud (or at least the flowering of one) that sprouts on the combined body of its parents, the fertilised egg.

Germinally understood, reproduction undermines the closed individuality of the organism from behind by opening it onto its generating conditions. Since those conditions are continuous through it, the organism is less a thing of its own than a derivation from the material of its progenitors. The germ-plasm is continuous not only through the processes of fertilisation and development responsible for the formation of the adult organism, but through its entire phylogenetic lineage as well. This continuity must be what Bergson has in mind when he claims that, in the attempt to locate the principle of the beginning of an organism, 'gradually we shall be carried further and further back, up to the individual's remotest ancestors: we shall find him solidary with each of them, solidary with that little mass of protoplasmic jelly which is probably at the root of the genealogical tree of life' (CE 43). The parents on whose combined body the bud of the organism first sprouted are each the flowers of budded parts of a combination of parental substances of their own. The process of their formation was determined and directed by germ-plasm that was continuous through them as well. The principle of closure that would secure the determinacy of the organism as a distinct individual is deferred backwards through each generation, arriving finally at the last common ancestor shared by all extant forms of life, Bergson's 'little mass of protoplasmic jelly' (CE 43).

'Where, then', he asks, 'does the vital principle of the individual begin or end?' (CE 43). The question is unanswerable, for 'each individual may be said to remain united with the totality of living beings by invisible bonds' (CE 43). It follows from this germinal unity that 'the individual is not sufficiently independent, not sufficiently cut off from other things, for us to allow it a "vital principle" of its own' (CE 42). The attempt to individuate the vital principle collapses back onto the entire history of evolution and encompasses 'the whole of life in a single indivisible embrace' (CE 43). The vital principle is a variant of internal finality. When Bergson claims that the individual is not sufficiently independent for it to be afforded a vital principle of its own, he means too that the individual is not sufficiently independent for it to be afforded a *final cause*, or inner purpose, of its own either. I return to this criticism of vital principles in chapters 5 and 6.

The closure required for the bounded individuality of the organism is frustrated by its reproduction and development just as it is by the relative mereological autonomy of its parts. Unifying these considerations is the same implicit premise. Internalism stands or falls with the complete reality of organic individuation. From this it follows, first, that in the absence of the ability to fully specify a limit between one living being and another – to individuate them – there can be no determinate distinction according to which purposiveness can be rendered internal. Second, and as a consequence, if finalism is to be viably thought at all, it will have to be rendered *external* to any ostensibly individuated biological form.

Here is the argument in schematic form:

1. Inner purposiveness requires determinate individuality
2. There is no such individuality in the organic domain
    2.1. Relative autonomy of organic parts from the individual whole
    2.1.1. Immunological challenge to organic self-identity
    2.2. The individual is derivative on its germ line
    2.2.1. It is impossible to determine where one organism ends and another begins
3. Thus, there is no way to locate finality *internal* to any biological individual
4. Thus, if there is to be finality, it must be *external* to all biological individuals

The idea that finality is a necessarily external attribution means that it is to be attributed to all of life indivisibly. Internality requires divisions to be cut into the organic domain, and since they are always incompletely determinate, the locus of finality will always reside external to them, on the outer side of any individual so considered. At the limit, this externalism is a global phenomenon, qualifying the whole from out of which internalism attempts to dissociate its individual parts. With this suggestion we have retraced our steps back to the first tenet of Leibnizian finalism, globality. Having reconstructed Bergson's argument against internalism, I turn below to the more popular critique of global finalism. That is also where I locate the relevance of Bergson's comments on the metaphysics of possibility.

b. The metaphysics of possibility

Bergson says that 'we have to choose between the out-and-out negation of finality and the hypothesis which co-ordinates not only the parts of an organism with the organism itself [internalism], but also each living being with the collective whole of all others' (CE 43). Internalism is undermined by the incompletable nature of organic individuality. Organisation and purpose cannot be internally generated features of the individual living being. They have to be imposed from outside of each, externally. Bergson discerns in their collective coordination the effectuation of a pre-existent model. If life is a globally harmonious phenomenon, that must be because it realises a programme set out for it in advance. This claim is intended both as a characterisation of Leibnizian finalism and as an inference made from the conceptual resources available to external finality after the critique of its internalist alternative. Bergson maintains that finality so construed is led back to the same actualism at work in the mechanistic explanations considered above and the position that 'all is given' in advance. Externalism merely employs a shift in the modal register of that 'all' – from the actual to the possible – without overcoming the abjuration of novelty that remains its necessary consequence. The 'actualism' at stake in external finalism is, counterintuitively, a possibilism instead.

Possibilism, at its most general, is the position that there are things that are not actual, things that could be actual, given the right set of causal conditions, but that need not be actualised. For Bergson, possibilism is the position that entities or events are possible before becoming actual. This is the modal metaphysics that Bergson discerns behind external finality. Ends or purposes are represented as possibilities that reside outside each individual and become actualised in the organisation of each individual. In this section I reconstruct Bergson's critique of possibilism in two steps, beginning with the falsity of the problem of possibilism, that is, the confusion of the less and the more; and then turning to the retrospective illusion that explains possibilism's persistent appeal.

i. The falsity of the problem

Bergson held that problems could be judged true or false independently of their solutions. A false or pseudo-problem is one whose terms are confused or badly composed. The idea of false problems was popularised by Deleuze in his lecture course and monograph on Bergson (2007: 84; 2006: 15–35). It has been a consistent theme in commentaries on Bergson since. The idea was apparently decisive for Bergson, who contended that 'the great metaphysical problems are in general badly stated, that they frequently resolve themselves of their own accord when correctly stated, or else are problems formulated in terms of illusion which disappear as soon as the terms of the formula are more closely examined' (CM 77).

Any attempt to solve such problems will involve a mistake because there is nothing to solve. If conceptual analysis cannot dispel the problem for good, this may not be because of the problem's profundity. Some problems retain their grip on the intellect as a result of the overextension of a pragmatic habit, such as the tendency to spatialise matter or to treat time discretely. False problems of this sort are not just errors, like incorrect calculations, but indices of the way human thinking evolved. In Bergson's words, there may be something like an intellectual 'illusion' responsible for their consistent misformulation. The proper treatment of such problems requires consideration of the relevant illusion.

Possibilism is a response to the problem of creativity falsely stated. The problem of creativity is the question of how it is that creative events – whether evolutionary, artistic, or otherwise – are to be conceived and explained. The false formulation of the problem asks how it is that new events come to be. The answer, for possibilism, consists in the supposition that 'the possibility of things precedes their existence', which means that 'they could be thought of before being realized [or actualised]' and that the creativity of an event, or the coming-into-being of an actuality, consists in the actualisation of some pre-existent possibility (CM 81). If possibilities are thinkable before they are actualised it is because they are real without being spatiotemporally determinate realities. They are ideal first, 'pre-exist[ent] under the form of an idea' (CM 83). Actuality is accomplished through the addition of concrete existence to the 'ideas' that prefigure it. Possibilism shares its basic precepts with Kantian artifactualism. There is less in the possible than in the actual; the actual adds something – that is, existence – to what precedes it, the way a building adds bricks and mortar to the blueprint on whose basis it is built.

The conceptual priority between the 'less' and the 'more' is temporally ordered. The series of actualities that constitutes the progressive realisation of the visual apparatus across history is to be modelled on this account in terms of a set of preformed possibilities that come into being by being actualised – which means nothing more than *located* – in space and time, in a set of individual bodies. Novelty is recast as the appearance in the actual of an entity that existed already in another modal register. If purposes reside outside of or external to particular individuals it is ultimately because they reside outside of or external to the modal register in which those individuals are registered as determinate actualities. To the extent that the possible precedes the actual, time adds nothing to and plays no positive role in the latter's realisation of the pre-existent model of the former.

Bergson's metaphysical ratification of the positive givenness of time, which registers as an effective flow of unforeseeable novelty, requires an inversion of the temporal order of possibilities. He contends that a truly creative evolution 'creates, as it goes on, not only the forms of life, but the ideas that will enable the intellect to understand it, the terms which will serve to express it' (CE 103). This inversion between the possible (ideas) and the actual (forms of life) is a consequence of the conceptual inversion between the 'less' and the 'more'. There is not less in the possible than in the actual. The actual does not come to be through addition, in this case of concrete existence to the preformed ideality of the possible. The opposite is the case. There is, for Bergson, more and not less in possibility than in the actuality that corresponds to it.

There is more to be found in the idea of possibility than in the actuality that it is supposed to prefigure because 'the possible is only the real [that is, the actual] with the addition of an act of mind which throws its image back into the past, once it has been enacted' (CM 81). The possible does not prefigure the actual; it is the actual that comes first. The ontological reality of possibility is to be replaced with the representational reality of an intellectual postulate whose ground is the actual entity that it purports to precede. Possibilities are ideas, but they are ideas that we form. We form them on the basis of the actual entities that they resemble. The possible/actual dyad should be reordered: it is possibility that comes to be through the addition of something more to what prefigures it.

Possibility is furnished through the addition of an intellectual operation of *subtraction*, coupled with the backwards projection of its result. Possibility subtracts from actuality precisely what the actual is supposed to add to it, its existence. So denuded, this idea or 'image' of the actual, which resembles the actual under a different modal category, is then projected backwards in time and located prior to its existence as actuality. It is this now ostensibly pre-existent image that is called the possible. The temporal priority of the possible thus established, the actual of which it was initially the image is conceived finally as its concrete realisation, attained through the addition of existence. Possibilism errs in confusing an effect for the cause of its cause. Its images are the effects of a series of intellectual operations performed on actuality. After projecting the effects of those operations behind the actual as actuality's pre-existent form, possibilism concludes that it is the actual that was subsequently realised on their basis.

ii. The retrospective illusion

False problems testify to ineradicable features of pragmatic intelligence triggered out of school, inappropriately extended beyond their original context. Possibilism is indexed to the 'simple law' of mechanism: '*the present contains nothing more than the past, and what is found in the effect was already in the cause*' (CE 14). The actual present is the effect, for mechanism, of a calculable rearrangement of the elements that preceded it in the past. For possibilism, it is the realisation not of pre-existent actual elements, but of preformed possibilities. Just as we are natural-born mechanists when it comes to matter, Bergson suggests that we are natural-born possibilists, or finalists, when it comes to life. But as the phenomena of life and matter differ in their presentation, a principle is required that would secure them together under the same intellectual law. Both assume that all is given in advance. Mechanism rationalises itself by deriving its intellectual habits from the tendencies of unorganised matter. On what basis does possibilism retain its appeal?

Here is where Bergson finds the 'retrospective illusion' at work (CM 82). The illusion involves the 'mirage' of the present in the past, of the actual in the pre-existent form of the possible. It is an illusion because only from the point of view of some present actuality can one perceive and conceive the antecedent possibilities realised in it. The retrospective illusion consists in the ascription of finality after-the-fact. From the present perspective of a completed process one looks backward over the events that are supposed to have culminated in its present phase as if drawn inexorably towards their telos. After the invention of the photographic camera, for instance, it looks as if the *camera obscura* and Daguerreotype comprised two phases in one and the same process, the end of which was always film photography. Of course, no such end existed for the calotype, as it is only after the film camera comes to be that one can locate its possibility in what look now to be antecedent conditions. One traces the outlines of a teleological process only in retrospect, by looking back from the perspective of its apparent terminus and selecting, distributing, and aligning the formative causes for its progressive realisation. This illusion is 'natural' to the intellect. The intellect is most at home in the causal-closure of unorganised matter; turning, then, to contingent and unforeseeable events in the history of life, and retaining a commitment to the idea that effects are always and completely contained in their causal conditions, the intellect is forced to carry out 'backwards over the course of time a constant remodeling of the past by the present, of the cause by the effect' (CM 84–5).

Possibilities exist, however, not only as mistaken inferences drawn from accomplished actualities by an intellect habituated on inert matter. Possibilities exist first and in principle as actuality's alternative-modal correlates, generated alongside the course of actual events and mirroring them in another register. This is what Bergson means by the pronouncement that 'as reality is created as something unforeseeable and new, its image is reflected behind it into the indefinite past' (CM 82). Reality 'finds that it has from all time been possible', Bergson explains, 'but it is at this precise moment that it *begins* to have been always possible' (CM 82, emphasis mine). Its possibility does not precede its actuality, but it '*will have* preceded it once the reality [that is, the actuality] has appeared' (CM 82, emphasis mine). The possible is a category with an extension identical to that of the whole past of the actual. It is a record of all accomplished actualities, testifying to the fact of their existence, like a photoreceptive plate in front of which the whole history of the world unfolds and on which that history is registered. Possibilities tell us nothing about the future (actualities-to-come). They tell us only about what has already happened (actualities accomplished), adding that what has happened actually is as a consequence subsequently possible as well. The possible is an essentially derivative modal category.

Counter-factuality – or the appearance of a possibility that has not been actualised – is a function of the correlation between two actualities that correspond according to some practical criterion when one has been emptied of its reality and reformed as a possibility. The contention, for instance, that it was possible for a colleague not to come into work today, even though she did, relies on the previous actuality of her staying home. I subtract reality from that past experience in order to arrive at its status as possibility, then I adjoin it with the current actuality in order to generate the counter-factual that she could have stayed home but did not. The point is the same. Beneath all possibilities – whether actualised or not – are the actualities from which they ultimately derive.

The illusion arises out of a suppression of this retrospectivity, of the fact that the possible is an effect of the actual, not the actual's antecedent cause. In order to overcome this illusion, the derivative status of possibilities has to be affirmed as a creative dimension of the actualisation process. This means, to take up one of Bergson's examples, that Shakespeare created not only *Hamlet* as an actual work of art, but the *possibility* of that artwork as well. *Hamlet* is possible only because and as an effect of its actual execution. It is possible, moreover, for all time, but it *began* to have always been so only as Shakespeare actually wrote it. Possibilities are timeless, yet they are generated from within time, as atemporal effects of the temporal unfolding of actuality. The atemporal status of possibility is what is at stake in possibilism's abjuration of the reality of time. The photoreceptive eye was on the possibilist account *always* (timelessly) possible. It is on that basis that the progressive actualisation of the eye over an array of phylogenetic lines and their convergence on similar phases of the same organ can be explained.

Here is the misstep. Possibilities have always existed, but only as retrospective effects of the actualities that they resemble. Their timelessness is produced from within time. They are inert, both in the sense that they do not change, that nothing new comes from them, and in the sense that they do not participate in the causal order of the actual world. They are its effects, its epiphenomena, mirroring its course without being able to influence it. The retrospective illusion passes over not only the temporal ground for the atemporal nature of possibility but possibility's epiphenomenality as well. Prey to that illusion, possibilism confuses the relationship between actual and possible and contends, as a consequence, that actual events are preceded by the possibilities realised in them. The possibilist undercuts the temporal reality of the actual world by locating beneath it the static programme of possibilities that structure its development. This, finally, is what it means to say that for possibilism everything is given in advance. Being is, as possible, not in time at all and therefore prior in principle to whatever comes to be actualised over the course of time.

The appearance of novel actualities is reducible, on possibilism, 'to a mere rearrangement of former elements', possibilities (CM 85). If these elements exist prior to their realisation in the actual, then they must be regarded 'as calculable and foreseeable', just as if they were the parts of an unorganised material system (CM 85). If a sufficiently powerful intellect could have predicted the history of the mechanist's world, then such an intellect could have predicted the course of the possibilist's as well. Whether it is the interaction of parts or the realisation of ideas that is at issue, in both cases what looks like novelty is the complicated rearrangement of things already given.

The alternative is that actual events have to be thought as unforeseeable in advance. This is not to say that their emergence and constitution has to remain inexplicable. On the contrary, actualities can always be explained, but only retrospectively. Only, again, from the perspective of the actuality already accomplished can one select and arrange its antecedent conditions in order to furnish a causal account for its emergence. Explanation is, at least in the domain of life, a necessarily backwards-looking affair. It should be held apart from its future-oriented counterpart, prediction – as Paola Marrati has suggested (2010). The appearance of the eye can be explained through definite causes, but this explanation after-the-fact cannot license the inference that what can be explained could have been predicted. The false equivalence between the two is exemplary of the retrospective illusion.

Genetics

Though Bergson's critique of possibilism has the genesis of teleological judgments and their metaphysical implications as its primary objects, the critique continues to bear on the core tenets of the genetic reductionism that has since come to define the study of development and evolution. This paradigm exists at the interface of mechanism and finalism, drawing precepts from both. It is not uncommon to find in reductionist embryology, for instance, an updated preformationism, or possibilism, of the adult organism in its genetic blueprint, or to read in a mechanistic theory of evolutionary dynamics a finalist appeal to 'possibility spaces' that structure the selection of beneficial mutations. Gene theory dissociates the metaphysics of possibility from finalism, unbinding possibilism from the principle of resemblance according to which possibilities are just actualities before actualisation and relocating it in an informational register as genetic code. Codes are supposed to precede and determine the development and evolution of the living beings in which they are replicated and across which they are transferred, but they neither resemble the living beings that they determine, nor do they require anything other than a set of explanatory mechanisms in order to account for their operation. I divide the last two parts of this chapter accordingly, beginning with the genetic account of development and concluding with the mechanistic account of evolution that corresponds to it.

a. Development

Few biologists have done more for the critique of genetic reductionism in the explanation of development than Richard Lewontin. To 'reduce' organism-level physiology to the cellular level is to explain the features of the former in terms of the vocabulary of the latter such that there is nothing remaining in the organismic register unexplained by its cells. Genetic reductionism is the position that every register in the study of life (ecological, organismic, cellular, protein, etc.) can be explained in terms of gene function. To the extent that genes work as mechanisms, genetic reductionism purports to explain all of life mechanistically (see Butterfield 2011).

While Lewontin's insistence that development should be understood across the myriad dimensions of the organism-environmental system might sound almost truistic today, it remains the case that many contemporary research programmes in the life sciences operate on the basis of a reductionism all the more trenchant for the increasing sophistication of the techniques and technologies through which it is realised. It comes as no surprise that the recent reaction against the reductionism of developmental biology in the wake of gene theory remains an object of scepticism among many biologists and a number of philosophers of biology as well. I think that there is a continuity between a certain tendency that still typifies gene theory and the commitments that Bergson identified across the mechanism and finalism of his time. Bergson can be

positioned in line with contemporary responses to gene-centric understandings of life.

Lewontin argues that the reductionism that has come to define contemporary biology was originally attendant on the conceptual vocabulary of a mechanistic understanding of nature. That understanding is
now in the process of being reconfigured in tandem with the development of an increasingly powerful set of techniques and technologies,
from recombinant DNA to molecular imaging. New understandings of
life have become coupled to a new series of mechanistic metaphors. It
is no accident that these metaphors are found informing developmental
biology, the field of study that takes as its orienting question how it is
that individual organisms begin as single cells and through processes of
division and differentiation come to assume the intricate complexity and
unique structural organisation that we associate with life. The beginning
of its answer is given in the name – development. Organisms are supposed to develop from out of the simple and similar into the complex
and different. But this is just a metaphor, even if a powerful one, and
it comes with its share of theoretical baggage. It is here that Bergson's
criticisms of mechanism and finalism – especially in the possibilist pre-
sumptions of the latter – can be seen to come together in updated form.

'Development', Lewontin explains, 'is literally an unfolding or
unrolling of something that is already present and in some way preformed' (2000: 5). This is a claim about the term 'development' itself,
but it can be taken in the spirit of Bergson's methodology of tendency-
analysis. At the conceptual limit of developmental biology's investment
in genetic determination is an understanding of change that biologists
call 'preformationism', the premodern position that imagined the adult
organism to be contained in miniature in the sperm.

Preformationism is a variant of actualism. Textbook histories teach
that it lost out to epigenesis, a theory of how the cellular content of the
fertilised egg undergoes changes in form over the course of embryogenesis in order to produce the matured organism. Today Hartsoeker's
infamous illustration of a tiny man curled up inside the sperm cell seems
long-since obsolete. But – in Bergsonian spirit – Lewontin contends
that it is in fact the preformationist theory of change that has hardened
into contemporary dogma. It survives in the metaphor we use to explain
ontogeny, and that metaphor organises approaches to the whole gene-
organism-environment complex. Philosopher of science Susan Oyama
agrees, arguing that the ontogenetic process responsible for the production of the fully formed organism is at the same time productive of the
'information' that would seem to determine it from the outset (2000: 4,
13–14). It only makes sense to invoke a set of genes that would 'initiate'

development on the basis of an arbitrary selection of where to begin the analysis (2000: 40). This may be the 'principle' of developmental biology's tendency towards reductionism, even if it is only ever approximated in practice. The assumption is that the ends of processes of development are in some way prefigured at the outset in identifiable ways, reducing development to the mere realisation of those ends across some stretch of time.

The development of the organism is usually explained in terms of genes and cell organelles. The environment is relegated to the role of background. The organism's genes are imagined to determine its final state, while the environment is conceived as a set of static conditions that either enable or constrain the way the genes express themselves. This view uses 'norms of reaction', which describe the pattern of phenotypic expression of a single genotype across different environments, as a baseline for determining the correlation between genes and traits. Relative to the 'norm', environments are not supposed to play causally efficacious roles in the production of organic form; they are supposed either to stabilise gene action or interfere with it by modifying its activity (see Oyama 2000: 180). There is little if any substantive difference between the idea that the organism is preformed in the fertilised egg and the more recent view that the complete blueprint of the organism as well as the information necessary for its organisation is present already in the embryo.

The genetic blueprint has updated the model of the miniature organism without modifying its basic architecture. Even though it presents itself under a mechanistic explanatory paradigm, it nonetheless borrows its sense from the external theory of finality according to which everything is given in advance in the form of a plan to be carried out. The genetic reformulation of external finality is based in a mechanistic principle of causal antecedence. It is not the intentions of a minded designer that condition the production of the organism but the blind mechanics of gene action that control for the organism's development.

Prior to the question of how genes act to control for development is the question of what it is that they do. The gene has been conceived from its inception as a causal agent, as Evelyn Fox-Keller has argued, long before the nature of its agency or its operation in developmental processes could have been specified (2000: 52). The first widely accepted account of gene agency grew out of the integration of the vague notion of 'gene action', popularised by Alfred Sturtevant (1932: 304), with Beadle and Tatum's formulation of the one gene–one enzyme hypothesis in the early 1940s. Having succeeded in linking specific mutations in *Neurospora* (a fungus) with changes in its metabolic pathways, Beadle and

Tatum claimed that genes control biochemical reactions, without yet knowing how. Then came Watson and Crick's work on the identification of genetic material with DNA and the articulation of the double helical structure in 1953. After that, a direct correlation was hypothesised between the sequence of nucleotides in a gene and the sequence of amino acids in a protein (the one gene–one enzyme hypothesis). By 1957, Crick had formulated a working hypothesis for what is today considered to be 'The Central Dogma' in gene theory, that a strand of DNA is specified by the sequence of its bases, and that this sequence codes for the amino acids in a particular protein (1958: 152). DNA makes RNA, RNA makes protein, and proteins make the organism. A few years later, Jacob and Monod introduced a distinction between 'structural genes' and 'regulator genes' to account for the fact that chromosomes seem to house both genes that code for proteins and genes that regulate the rate at which that coding can take place. They shifted the discourse from talk of 'gene action' to 'gene expression', substituting for the metaphor of genes as 'blueprints' for development the subtler metaphor of the organism as the realisation of a genetic 'program' written into its DNA (see Jacob and Monod 1961: 354).

The state of the cell regulates this program. It is responsible for running the transcriptive-translative processes of editing and splicing in some determinate fashion. The locus of control is relocated from DNA (structural and regulatory genes) to the cellular system in which DNA performs its function. The processes of editing and splicing 'form' the gene as a functional part of the system from the gene as a pre-determinate string of DNA, separating structural from functional genes. It is as if it is the function of the gene (as end-product of the transcription processes) that determines its structure (as inherited contiguous string of DNA). The function of the protein to be formed seems to control for the formation of the mature mRNA transcript. The structure of the mature mRNA transcript, the structure of the functional gene, seems to be determined by the function of the protein to be formed. It may not be going too far to discern here a surreptitious teleology.

Jacob and Monod would eventually call a more sophisticated version of this new account the 'operon model', referring to the coordination of a network of regulatory elements and structural genes by a regulator gene situated elsewhere in the genome. The regulator acts by controlling for the synthesis of a repressor that binds to an 'operator' region adjacent to the structural genes and regulating their transcription. According to the Central Dogma, transcription-translation proceeds through the following phases: (1) DNA is separated in the nucleus (its strands are dissociated); (2) *transcription*: RNA polymerase copies the bottom strand in complementary messenger RNA (mRNA); (3) the mRNA is transported from the nucleus into the cytoplasm; (4) the transfer RNA (tRNA) binds to the mRNA on the ribosome by recognising the mRNA's triplet codons; (5) *translation*: the tRNA adds to the protein polymer chain an amino acid monomer; finally (6) the protein chain drops the ribosome and performs its function in the cell. It was eventually discovered, however, that complex splicing and editing processes intervene between (3) and (4) because proteins occasionally require separated strands of DNA for their synthesis. Often there are strands of inactive or 'junk' DNA ('introns') between the strings of active DNA ('exons'), so the introns have to be edited out from the primary mRNA transcript and the exons have to be spliced to form the mature mRNA transcript. Exons can be spliced in different orders. Unexpressed DNA can evolve by drift, or mutation, more or less independently from natural selection (which works only on phenotypes). More than one mature mRNA transcript is possible as a result of the same primary mRNA transcript. There can be no direct correspondence of DNA to synthesised protein, and the Central Dogma fails on this central count.

The genetic program is one more variation on the same theme that organisms are the outward realisation of sets of causally determining inner mechanisms. The standard method for demonstrating this kind of determinism is as follows. First, isolate and identify a genetic mutation that prevents the organism from manifesting some regular trait, as in the case of a mutation in *Drosophila* that prevents its wings from forming. Then infer from the fact that a genetic abnormality obstructs the development of the trait to the idea that the relevant gene accounts for its normal development. Finally, extrapolate from that causal relation the more general claim that each of the organism's traits are determined in like fashion by its genes. Following Lewontin, ontogeny is best understood as the consequence not only of relevant genes, but of the interaction between those genes and the character of the external environments through which the organism passes and the sequence in which it passes through them. What Lewontin calls 'random events of molecular interactions within individual cells' play a role as well, the effects of which can never be specified in advance and have little to do with the organism's genetic makeup (2000: 17–18).

Wedded as it is to the claim that the organism manifests an inner program, reductionism tends to eschew the causal efficacy of environmental conditions. Lewontin considers the case of the tropical vine *Syngonium*, whose morphology fluctuates according to the light conditions in which it develops. This vine provides an apt example of two facts. The first is that stages of development vary relative to the environment. The second is that the identification of 'stages' is a somewhat arbitrary affair, since organisms develop in tandem with both internal and external changes.

There is no way to predict the outcome of a developmental process in advance or independently of a unique sequence of salient environmental factors. The same point holds evolutionarily. We cannot know what will be favoured by natural selection without a specification of the history of environments each species will meet in the course of its evolution.

It has been said that the advent of adaptationism marks a rupture in the study of life by alienating the interior of the organism from its exterior, separating internal processes responsible for individual variations from the external pressures that determine which are retained and eliminated at a population level (see Levins and Lewontin 1985). This separation reaches its apogee in the causal independence afforded to the environment, which is supposed to interact with the organisms that populate it only through the interface of the selection process. The organism endeavours to adapt itself to the conditions of its environment and either fails or succeeds, while the environment fluctuates according to an independent set of climactic and geological variables. As biology meets computer modelling and mathematics, the organism itself – its unity, morphology, and behaviour – is dissolved into a computational algorithmics of allele frequencies distributed across fitness landscapes.

While this kind of vulgar adaptationism still has its proponents, the question of the status and efficacy of the whole organism in evolution is exciting renewed theoretical interest (see West-Eberhard 2003 and Lamb and Jablonka 2005). Brian Goodwin (1994) was among the first to herald this shift back towards the organism in evolutionary thinking by enlisting the recent findings of the complexity sciences and their revitalisation of the concepts of emergence and downward causation. This shift has yet to inspire a move away from metaphors that reduce the organism to its genes and background the environment as stable, static, and constraining, though that too appears to be changing. Bergson may have been able to foresee the general thrust and broad outline of at least some of these developments. His critical engagement with the reigning biological paradigms of his time continues to track many of their core philosophical presuppositions through to today. While there is scarcely more reductively mechanistic an approach to life than the genetic, possibilism is retained in its application to the study of development. Gene theory may be susceptible not only to Bergson's critique of mechanism but to his critique of the metaphysics of possibility as well.

Genetic possibilism culminates in Richard Dawkins' declaration that the gene is to be regarded as potentially immortal (1989: 24, 34–5). Dawkins establishes this potential immortality by radicalising the Weismann barrier and arguing for an absolute diremption of biological organisms from the genetic information encoded within them. Organisms are better understood as the 'vehicles' or 'survival machines' in which genes are replicated and distributed. Vehicles die, while the information on the basis of which they are built exists only in its copies and is therefore potentially immortal. As genes become informational, possibilities can be seen to recuperate the atemporal nature that is, for Bergson, their strictly retrospective essence. By pulling the gene up to its hyperbolic apex in the immortality of replication, Dawkins reveals the possibilism that resides at the core of the genetic account of development.

There is, of course, more to reproduction than replication, and the genetic information replicated is inconceivable apart from the cellular networks in which it is instantiated – DNA is only maintained in a persistent state by a machinery of editing and correcting processes in the cell – as well as the external milieu in which the internal systems develop. Dawkins is simplifying away a lot of crucial biological content in order to make a point. It is difficult to see how the effect of any determinate gene qua replicator on the phenotype of a given organism could be isolated and targeted by selection. Evolutionary theorists have elaborated this problem in terms of 'hitch-hiker' genes. Selection for a certain phenotype will always involve the selection of genes correlated with that phenotype that do not causally contribute to its realisation. These genes can be said to 'hitch-hike' along with the others that appear to be targeted. There may be a lot of genetic variation that is 'invisible' to selection.

b. Evolution

Daniel Dennett's *Darwin's Dangerous Idea* elaborates a theory of evolution by natural selection that exemplifies the striation of the genetic explanation of life across Bergson's mechanist/finalist alternative. For the most part, I follow Keith Ansell-Pearson's critique of Dennett in this section (see 2002: 79–86), though my emphasis is different. Instead of arguing that Dennett subtracts the novelty from evolution – I agree that he does – I argue that the subtraction operates according to a cross-over between mechanism and finalism that finds its basis in genetic reductionism, and I think Bergson can be brought to bear on that register as well.

The mechanism of Dennett's account is furnished by his reconception of Darwin's theory of natural selection as an 'algorithm', 'a certain sort of formal process that can be counted on – logically – to yield a certain sort of result whenever it is "run" or instantiated' (1995: 50). Algorithms are defined by three basic features: substrate neutrality, underlying mindlessness, and guaranteed results. Algorithms are substrate neutral when they can be reliably instantiated in any medium whose causal powers permit the algorithm's procedure to be carried out. They evidence an 'underlying mindlessness' when each of their constituent steps and the transitions between them are simple enough to be executed by a basic mechanism. Algorithms, despite producing potentially brilliant results, consist of nothing besides a series of steps basic enough to be run by a rudimentary mechanism. No mind is required, to use Dennett's words, in order for the algorithm to work; it is '"automatic" by definition: the workings of an automaton' (1995: 59). If each of its steps is properly executed, then the algorithm produces the same results every time. Darwin's so-called 'dangerous idea' is the position that evolution is an algorithmic process of just this sort.

Dennett does not specify the steps that are supposed to constitute the algorithm of natural selection. It is more important for him to establish that the results of the evolutionary process can be explained algorithmically. Here is a reconstruction of those missing steps:

1. Organisms inherit characteristics from their parents, but also depart and differ in small ways from them: they vary.
2. Organisms produce more offspring than can viably survive.
3. Descendants that differ in a beneficial way have a better chance of surviving and reproducing than those typified by harmful or otherwise non-beneficial variations.
4. Since the offspring produced by benefited organisms will themselves both resemble and differ from their progenitors, the process repeats itself. Return to step 1.

Natural selection works by coupling variation to selection in order to produce design. This process is supposed to be substrate neutral, mindless, and reliable. Wherever there is variation over time and something to secure the replication and conservation of some variations over others, there is natural selection at work. Run it for long enough in the medium of biological life forms, and it will account for the gradual emergence of every innovation accomplished in the evolution of life.

Dennett's Darwinism retains the notion of design often employed against evolutionary theory by detaching the explanation for design from the minded intentionality of a designer and reconfiguring it in terms of the mindless automaticity of an algorithm. 'Design', as Dennett has it, 'is Aristotle's *telos*, the exploitation of Order – mere regularity, mere pattern – for a purpose, such as we see in a cleverly designed artifact' (1995: 64). While the solar system evidences an order, a patterning captured in mathematical models, it does not appear to have a purpose. Its order is not leveraged to accomplish anything, so it does not exhibit what Dennett calls design. An eye exhibits design. It has a function, seeing, and its function is only possible through the exploitation of order, the structure of the visual apparatus. Dennett understands Darwin to have explained the production of design from out of nothing but order over time. 'Let me start with regularity', says Dennett's Darwin, 'and I will show you a process that eventually will yield products that exhibit not just regularity but purposive design' (1995: 65).

Design is explained through the enlistment of 'cranes' or through the postulation of 'skyhooks'. A 'skyhook', as Dennett defines it, 'is a "mind-first" force or power or process, an exception to the principle that all design, and apparent design, is ultimately the result of mindless, motiveless mechanicity' (1995: 76). Any explanation of the appearance of the eye that involves a principle other than natural selection has to appeal to a 'skyhook', a transcendent interruption of the mechanistic order that intervenes on that order's causal processes from the outside, a *deus ex machina* of evolutionary theory. 'Cranes', by contrast, operate by 'lifting', which means that they do not enlist in their explanations anything other than the basic mechanical processes already at work beneath them to account for a given creative phenomenon or evolutionary event. To explain the evolution of the eye by way of a crane is to account for it from the bottom up, through the algorithmic exploitation of order for purpose fed back into itself recursively – using small cranes in order to build bigger ones – to produce ever more complex structures with ever more sophisticated purposive specifications.

Though neither cranes nor skyhooks may be foreseeable in advance, cranes can be demonstrated to have been predictable in that they are explicable after the fact through nothing other than the same mechanistic processes that preceded them. Skyhooks remain unpredictable even retrospectively; they are explanations that require the supplementation of the mechanistic processes that preceded them with additional principles. Keith Ansell-Pearson suggests that, if it were to be accepted, this binary explanatory alternative would require us to dismiss everything from Nietzsche's will to power to Bergson's *élan vital* as skyhooks (2002: 82). Though I agree that Dennett's philosophy of explanation is unduly constraining, I do not think that theoretical postulates like the will to power or the *élan* should count as skyhooks, for the simple reason that they do not intervene into an otherwise mechanistic order from the outside. They are metaphysical principles understood within each system to be co-originary with the material processes that they supplement. They are cranes, by definition. The departure they effect from the mechanistic order valorised by Dennett does not occur in an occasional interruption from the outside but in a reconfiguration of that order from within. For Bergson, mechanistically explicable systems are largely the result of an evolved intellectual habit. It would make little sense to conceive

Bergson's creative principle as an intervention into matter, if matter is to be understood on the basis of an intellect adapted to it.

It is no accident that Dennett relies on the spatial metaphors of 'lifting' (from beneath) and 'hooking' (from above) to explain achievements made in organic design. No sooner than he introduces the concept of the crane does he postulate a 'space' in which it is supposed to do its lifting. This 'Design Space' is a version of game theory's 'possibility space', a map of all available options for action laid out such that a series of actualities can be modelled as a trajectory taken through space. Richard Dawkins posits an analogue in *The Blind Watchmaker*, which he calls 'Biomorph Land' – another sign that genetic reductionism requires a finalist (or possibilist) supplement to function as an account of evolutionary creativity. Stuart Kauffman, writing in a less reductionistic vein (1995), draws on the same conceptual vocabulary to explain how and why it is that the actual protein combinations evidenced in extant biology comprise only a small subset of all the possibilities 'adjacent' to them. Genetic possibilism is not unique to Dennett, though he does provide it with an especially clear formulation.

Dennett's Design Space is a space of all possible phenotypes – or design plans – that can be built out of some combination of genes. He conceives the latter as a 'Library of Mendel', an analogue of Borges' Library of Babel, which would contain 'all possible genomes', or DNA sequences, in a genetic alphabet comprised of stable permutations of adenine, cytosine, thymine, and quinine (1995: 112). Though both are possibilist postulates, there is a difference between the space of possible genomes and the space of possible organisms, since

> the fact that *we* can consistently describe a finished product – say, a giraffe with green stripes instead of brown blotches – does not guarantee that there is a DNA recipe for making it. It may just be that, because of the peculiar requirements of development, there simply is no starting point in DNA that has such a giraffe as its destination. (1995: 117)

Design Space is nested within the Library of Mendel, one abstract level of possibility beneath it.

Dennett maintains that Design Space is the proper medium in which the algorithm of natural selection is instantiated, on which it operates, and through which it moves. Natural selection is an algorithm that actualises possible designs by searching for them through possibility space and constructing them out of available genetic 'recipes'. The reasons why will become clearer in a moment. Note for now that, whatever its rationale, the theory of Design Space involves the introduction of the metaphysics of possibility into an otherwise avowedly mechanistic programme. Dennett's mechanism is not substituted for a finalist conception of the creation of biological forms but only for the intentionality that would direct their actualisations. Dennett replaces the designer with an algorithm while retaining the language of design. It is difficult to imagine a better example of what is for Bergson an intellectualisation – or spatialisation – of evolutionary creativity. All possible phenotypes are set out in advance. Their progressive actualisation over evolutionary history is a result of a set of blind mechanisms set to work on a minimum of material order. It seems reasonable to suggest that Dennett's possibles, according to his own principle of differentiation, are closer to skyhooks than they are to cranes.

One abstract level down from phenotypic possibility is a space structured by 'forced moves', or biological necessities, that determine what a phylogenetic trajectory must do when faced with certain problems. Possibilities fan out around each move in cascading fashion, just as in a board game. Spatialising the process affords Dennett an answer to the question of why it is that the set of all actual genomes comprises only a 'Vanishingly small subset of the combinatorially possible genomes' (1995: 124). Some biological possibles were 'more possible' than others because 'they were *neighbors* of actual genomes, only a few choices away in the random zipping-up process that puts together the new DNA volume from the parent drafts' (1995: 125). Some possibilities will be nearer and 'easier to get to' than others further away from some given trajectory through the space (1995: 118–19). As evolutionary history unfolded, and actualised genomes 'began to move away from the locations in Design Space of the near-misses, their probability of ever happening grew smaller' (1995: 125). Design Space is dynamically weighted in favour of the repetition of actual determinations. As the same possibles are actualised again and again, they become increasingly *probable* as well. Possibility is spatialised and structured topologically, deformed around the actual paths already traced through it.

Within the space of possibilities constrained by 'forced moves' there is the 'good trick'. Good tricks are easy, successful ways of solving particular problems. As those problems arise across evolutionary history, we should expect to find the appearance of the same solutions in response. Good tricks tend to be hit upon more often than other options. Though there is no requirement that they are actualised, there is no *telos* orienting evolution towards them, they nonetheless represent something like a rational or rationally explicable solution. This claim elides Bergson's distinction between predictability and retrospective explanation. Even if some solutions appear to be rationally explicable after the fact in a way that other solutions do not, it is not therefore the case that those solutions were the more rational alternatives in advance. That would require that basic life forms, confronted with a series of solutions present as possibilities, could reason about whether to develop vision or to take some other available route. We can enumerate those routes and rank them in terms of their rationality only retrospectively. We cannot even foresee them in advance, let alone adjudicate among them.

Dennett thinks that 'good tricks' work well and are therefore 'like beacons in Design Space, discovered again and again, by the ultimately algorithmic search processes of natural selection' (1995: 144). Anywhere we find life forms of sufficient complexity, we should expect to find autonomous metabolic systems that enable organisms to take in fresh material from the outside, replenish themselves, and resist the thermodynamic tendency to break down (1995: 127). Metabolism is a 'good trick' – and a crane – for prolonging organic complexity in the face of thermodynamics, and will tend to be realised wherever those conditions obtain. The same explanation is supposed to suffice even in more sophisticated cases of convergence, like the eye. Add another set of conditions to the foregoing: locomotion and ambient light. Information about distal objects can be fed into the locomotive system in order to complexify it, and 'such information can be garnered in a high-fidelity, low-cost fashion by vision' (1995: 128). Dennett considers vision to be a 'good trick', actualised wherever possible. Eyes are not always available in Design Space – given forced moves and cascading possibilities – but whenever they are, we should expect locomoting organisms to evolve them.

Dennett unifies the mechanistic conception of the organism as an inert system whose parts can be analytically decomposed down to the genes responsible for constructing them with the finalist postulation of possibilities that resemble and pre-exist their actualisations over time. Gene-theory provides the basis for the unification. Instantiated in a spatialised medium of possible genomes that is specified in terms of phenotypic design plans, structured through 'forced moves' and punctuated with 'good tricks', the algorithm of natural selection suffices to explain the whole of evolutionary history on the basis of the combinatorial logic of genetics (1995: 143). If it is true that, in Dennett's own words, 'with hindsight, we can say that tigers were in fact possible all along, if distant and extremely improbable', it is because the genetic combinations required to construct each of their organs were available all along, awaiting the searching mechanism of natural selection in the space of biological possibility (1995: 119).

Dennett's possibilism operates retrospectively. Design Space amounts, in the end, to the reified postulate of an intellect habituated on inert matter and its recombinations. Dennett does hold that some features of evolutionary causality work retrospectively. Since accomplishments in design are always accumulated on previous design accomplishments, each new 'discovery' makes a contingent 'discovery' retrospectively necessary. DNA did not have to develop – to take Dennett's own example – but after it did, it allowed for the evolution of life as we know it now. Its necessity was conferred upon it only after evolution had begun to work on its basis. Dennett thinks the same for speciation events. It is only after they have taken place that we can look backwards and identify the event defining the emergence of a new species. This is not an epistemic constraint, as if we could eventually learn how to identify these events in advance. It is in the nature of evolution that it only confers a necessity on its reasons after the fact, since at the time of their appearance they were not yet defining reasons for what would come next. Dennett holds together these claims with his otherwise trenchant possibilism about evolutionary creativity by endorsing a distinction between 'reason' and 'cause'. It must be the case, since all possibilities precede their actualisations, that for every evolutionary event its cause necessarily preceded it as a (non-retrospective) possibility. But it does not follow that the *reason*, or determining factor, for every evolutionary event precedes it as a possibility. It may be only after one possibility is actualised on the basis of another that the first becomes the 'reason' for the second. This is how Dennett is able to attribute retrospectivity to evolutionary reasons while remaining a possibilist.

Dennett's mechanism operates on the assumptions made by spatialising intelligence. It is only by reducing the organism – developmentally embedded in a fluctuating environment that its own activities help constitute, opened onto contingent networks of relations with other organisms, and regulated internally by an ecological immunity – down to the static combination of the fixed symbols that comprise its genome, that it could be modelled as an artifact built, however mindlessly, on the basis of the design plans that pre-existed it.

Conclusion

The component moves in Bergson's critical engagements with mechanism and finalism can be updated according to some of the recent developments undergone by both. Bergson's critical programme continues to track contemporary, gene-based accounts of evolution, especially when they fall back on the actualist (or possibilist) assumptions that he had already identified and criticised in the mechanism and finalism of the late nineteenth and early twentieth centuries.

According to mechanism and its implication in adaptationism, evolutionary accomplishments are described by the coupling of random variation with environmental selection pressure, producing like solutions to like problems. The adaptationist explanation of convergence has recently been supplemented with a developmentalist account, but it continues to retain a set of actualist assumptions that render it inadequate to the reality of convergence. Finalism is to be criticised for similar reasons. Kant's inner purposiveness requires a conception of biological individuality that we do not find fully realised in living systems, and Leibniz's global plan is actualism inflated to subsume the modality of possibility as well.

These are the key critical facets of Bergson's philosophy of evolution. Now we can turn to the positive project, beginning with Bergson's own metaphysics of modality. If the primary object of Bergson's critical programme is actualism (and its possibilist variation), then his positive philosophy calls for an alternative modal supplement. In Chapter 2, I draw a conception of that modal category, called the virtual, from a reading of *Matter and Memory*. The rest of the book leverages the idea in service of reconstructing the constructive dimensions of Bergson's philosophy of biology.

2

The Virtual: Instantiation, Implication, Dynamics

Bergson theorises a type of existence distinct from actuality and possibility, called virtual. This chapter traces this category through a set of themes in Bergson's philosophy before its employment as the modal designation for the *élan vital* in *Creative Evolution*, which I examine in the next chapter. The themes I discuss in this chapter include Bergson's accounts of perception, behavioural schemata, the past, and intellectual effort. I use them to map a conception of the virtual, pausing to elaborate some consequences for the philosophy of biology as I proceed. The chapter concludes with a global definition of virtuality according to the mereology of instantiation, implication, and dynamic interaction.

Introduction

Bergson's philosophy of biology pivots on a modal category of virtual existence, a type of existence irreducible to the categories of actuality and possibility that I discussed in Chapter 1. The evolution, development, and organisation of living things are patterned by tendencies. Tendencies are not manifest all at once, in actuality, and neither can their trajectories be articulated in advance, like possibilities. Understanding them requires the use of a different philosophical vocabulary, a vocabulary of virtual existence intended to capture the features distinctive of life, such as its temporal dynamism and the interpenetration of parts in an organised whole. This chapter lays the details of that vocabulary out.

Gilles Deleuze was among the first to make much of anything serious of Bergson's talk of virtuality (1991: 96–7; 1994: 201–14; 2006: 51–72, 91–113). Almost all recent work done on the term refers to Deleuze. It is typically taken for granted that Bergson develops something like a theory of virtuality, though there is no consensus on what that theory entails or whether Bergson is consistent in his own employment of it. The four panellists at the Society for Phenomenology and Existential

Philosophy's 2017 'Bergson Circle', the topic for which was *Matter and Memory*, agreed that the text's usage of the term 'virtual' is inconsistent, operating on some occasions as a rough equivalent for potentiality, on others as a synonym for dormancy or uselessness, and on others as a specific designation for the reality of the past (Ansell-Pearson et al. 2017). I demonstrate that there is a deeper continuity at work, and that the various senses that Bergson accords to the idea of virtuality throughout *Matter and Memory* can be brought together. I hold in particular that the virtual is a modal category, and that it is best understood in mereological terms, as a theory of part–whole relations.

The virtual is not typically interpreted in this way, as a theory of mereology. Keith Ansell-Pearson is one exception. His view is that Bergson deploys the idea of virtuality in *Matter and Memory* as a designation for the primacy of a whole over its parts. On this reading, the whole is real while its parts are virtual and derivative upon it (2005a: 1112). I think that Ansell-Pearson is right to notice that virtuality has something to do with part–whole relations, but wrong to think that it refers to a derivative modal status that is particular to parts. First, the virtual is not opposed to the real (as Ansell-Pearson suggests) but to the actual; second, I argue that it is the virtuality of the whole that is contracted in the actuality of each of its parts.

Bergson uses the term in his discussions of perception, behaviour, and memory; he also refers to it in his study of intellectual effort, before enlisting it as the modal designation for the *élan vital* in *Creative Evolution*. The virtual appears first in *Matter and Memory* as a way to conceive action possibilities that are present in the objects upon which they can be performed. Objects of perception reflect what can be done with them. Possible actions are virtual in the objects of perception and actualised when executed. I call this instantiation. It is one facet of my interpretation of virtuality. Bergson also uses the term to refer to the way a behavioural whole can be contracted, as a whole, in the particular movements through which it is performed. Such a whole is virtual in each of its actual parts. I call this implication. It is the second facet of virtuality. In his discussion of memory, Bergson regards the past as a whole that is reconfigured by the addition of new images to it, as well as in each particular contraction of it. I call this dynamism, or dynamic interaction. I take it to be the third facet of virtuality. My view is that the virtual is a modal category, an alternative to possibility, defined by the mereology of instantiation, implication, and dynamic interaction. This chapter elaborates each in turn, showing how they work to explain more familiar ideas such as duration and memory, and pausing to consider their consequences for the philosophy of biology. I use virtuality to make sense of Bergson's *élan vital* in the next chapter.

Instantiation

a. Images

Matter and Memory marks the beginning of Bergson's considered use of the idea of virtuality – though some of its valences are brought forward from concepts worked out initially in *Time and Free Will*. The primary postulate of *Matter and Memory* is the image. It is one of Bergson's most important ideas. Images are like pictures, qualified things, delimited and determinate (see Lawlor 2003: 4–10). They are what make up the actual world of material objects, subjective percepts, and past recollections. The basic ontological unit of the given is an image of it. Images are what Bergson calls 'spatializations', a concept I return to in Chapter 4. A spatialisation is a fixing or determination in space of what is, considered temporally, an unfolding stream of activity in which each moment prolongs the last and fades into the next, like a melody.

While the material image of an atom is bounded – something we can picture at an instant – considered in time it spreads itself across a series of vibratory oscillations, influencing and influenced by the vibrations around it and ultimately continuous with them (MM 265; cf. Čapek 1971: 189–291). Bergson goes so far as to declare that *all* division of matter into independent bodies (or images) is, in the last analysis, an artificial operation (MM 259). The material world, independently of its perceptual articulation in the form of images, is a decentred aggregate of points – relative centres of force or influence – each one of which 'gathers and transmits the influences of all the points of the material universe' (MM 31). It is a continuous whole, an energetic flux. Bergson endorses the physicist Michael Faraday's reconception of the atom as a point of compression out of which lines of force emanate in every direction, ultimately shaping every other material point in the solar system (see Blanchard 2008). If the material thing *is* where it *acts*, then on this view, every atom fills the world (CE 203). The appearance of matter in the form of distinct bodies is the appearance of images.

Bergson's theory of the matter-image comes close to what scholars call field-metaphysics (see Bennett 1984). The field-metaphysical account of determinate, individuated entities understands them in terms of the way they occupy subregions of space. Subregions are conceivable only through the concept of the whole, or field, in which they are embedded. The field is continuous and unified; parts of space, or individual extended bodies, are not really distinct from it. Determinate bodies are alterations or modifications local to some subregion. They are like temporary agitations of the larger spatial field, distinct individuals perceptually but not ontologically so (see Viljanen 2007: 406). Bergson sounds field-metaphysical when he defines matter in terms of '*modifications*, *perturbations*, changes of *tension* or of *energy*, and nothing else' (MM 266). Matter is not really divided in itself; it is differentiated spatially, while 'the essential character of space is continuity' (MM 259). Matter is distributed across a field of which each determinate body comprises a subregion defined by change in tension or energy, articulated and localised there by the perceiving subject whose interests it serves.

Matter is continuous. Material images, or apparently distinct bodies, are the effects of a reduction of relational complexity and a cut into continuity. Perception effectuates the reduction and makes the cut. Perceptual objects, or images, are narrowed versions of larger, interrelated wholes, slices cut from wider fields and centred around privileged positions. They are not different in kind from matter. Bergson inverts the theory of secondary qualities. It is not the case that perception adorns its objects with sensory qualities that objects do not themselves possess, as on the classical view. Whitehead called that view the 'bifurcation of nature'. The distinction between primary and secondary qualities is a bifurcation: it sets bare, colourless matter apart from the minds that project onto the world the vibrancy, tone, and significance it has in experience. 'Nature', in Whitehead's terms, is consequently made into 'a dull affair, soundless, scentless, colourless; merely the hurrying of material, endlessly, meaninglessly' (1967: 54; cf. Debaise 2017: 5–15).

For Bergson, precisely the opposite is the case. Perception selects away from its objects, dividing a continuity of matter into distinct bodies and filtering it down to a set of images of interest (MM 28). That means that there is more in matter than what is given in perception, but not something other, not something necessarily hidden or ontologically distinct (MM 78). The operator of this difference – between the less and the more, the part and the whole – is the living body. Bergson defines it as a 'center of action' that introduces a 'zone of indetermination' into a system of material images (MM 5, 23).

The ability to act – instead of just to react and transmit energy or influence along some causal chain – is the index of a privileged position in a system of images, a centre around which those images are constellated. The relation of this centre to the images that surround it is causally indeterminate. Living bodies can react in a number of different ways to what stimulates them. That difference is the mark of life, striated across the whole evolution of living forms, from the almost mechanical behaviour of the protozoa to an almost complete interruption between stimulation and response in higher mammals.

Bergson theorises the phenomenon in terms of hesitation. An organism's nervous system and brain – as well as their functional analogies in other life forms – receive stimulation, transmit it to relevant motor centres, and present the organism with the largest possible number of responses available. The more possibilities there are for action, the less immediate the execution of any one of them is, and the more an organism can hesitate before reacting.

Hesitation is correlated with a certain perceptual scope. Bodies simple enough to have to react almost immediately to stimulation, in a way that is basically circumscribed in advance, seem to be incapable of perceiving much beyond the thresholds of what stimulates them (see Berthoz 2008). They have no need for larger perceptual fields. Perception is a function; it serves the purposes of action. The differentiation of sensory organs means the registration of a wider variety of influences and the ability to perform a larger number of possible actions in response. Hesitating means perceiving possibilities and considering implications at an increasingly wide range of extension. Bergson puts the point in the form of a law: '*perception is master of space in the exact measure in which action is master of time*' (MM 23). Action is master of time in the exact measure to which the living body has divided its physiological labour across specialised organs and systems (MM 28, 57).

b. Possibilities

The first facet of virtuality, instantiation, consists in the way the perceiving subject refers these perception-images back to itself (see Parmentier 2017). Perceptions are material images that reflect back to the body what it can do (MM 30). Those action possibilities, before they are performed, are virtually instantiated in the relevant perception-images. In the literature on contemporary mereology, instantiation refers to the presence of a property, usually a universal, in a particular entity – like the property 'greyness' in a grey object, a pen (Cowling 2014). Actions are present, or instantiated, in the objects upon which a given body can perform them. The ontological status of those actions is virtual. Virtuality names the modal register in which action possibilities are predicated of the perception-image. These possibilities do not exist solely in the mind of the perceiver, like projections, nor are they actual features of the material image itself, taken independently of its registration in some particular perception. They are in the image, instantiated there by the way it reflects a body's relation to it.

An organism sees its action possibilities in what it perceives, and the set of those possibilities works to filter the complexity of the world through what the body can do, that is, the 'eventual influence of the living being upon [it]' (MM 30). These virtual actions are instantiated throughout the perceptual field, structuring it in terms of its navigability, utility, uselessness, or danger, in something like the way the rules of a game organise a field of play in terms of how it is to be traversed, what is to be avoided and what sought after – or what can be done on it. Perception is a 'filter' or 'screen' for the world rather than anything added or superimposed upon it (MM 309, 32). Perception's bandwidth is outsized by the continuity of matter. The body lets through only those features of matter that practically concern it; it obscures everything else, like an optical lens that brings some area of the visual field into focus by blurring whatever surrounds it. Secondary qualities are the effects of diminution or attenuation. The difference between being and being perceived is a matter of mereology, a mereological difference. The world is not *other* than the way we perceive it, but it is more than that – a whole beyond its perceptual particularisations (MM 27, 78).

Perception does not generate images; it simplifies a whole into parts by filtering it through a sieve whose mesh consists of the body's possibilities for action. Those possibilities are graded: they consist first, and most widely, in the actions a given body is capable of performing physiologically; second, and more narrowly, in the actions that body will tend to perform, or the actions towards which it is predisposed; and third, most narrowly, in the learned behaviours that development and enculturation have installed. Perceptual filters – as one prominent Bergson scholar observes – are composed not only out of 'the physiological constraints of our organs of perception and our inbuilt or instinctive action schemata, but also by the kinds of thing we are able to do because we have learned to do them, and need or want to do them' (Moore 1996: 29). Thus, the taxonomy of possible actions: physiological, behavioural, and subjective. Each involves what a body can do, and together these capacities operate as a filter for the world. As Deleuze has it, 'perception is not the object *plus* something, but the object *minus* something, minus everything that does not interest us', that is, everything that does not relate to one of our capacities to act on it (2006: 25).

Perception, in other words, is not sensory but sensorimotor. Perceptions indicate the nascent bodily actions that will complete them (MM 92). To perceive is to be moved to act, and 'perception as a whole has its true and final explanation in the tendency of the body to movement' (MM 41). The perceptual registration of an object tells us what we can do with it. Philosophers of mind call this theory of perception 'actionist' or 'enactivist', in contrast to accounts that privilege sense-data, representation, the localisation of experience in the subject, or disjuncts between perception, action, and conceptualisation (see Noë 2012: 65). The world 'shows up', on the actionist account, '*as available*, as within reach' (Noë 2012: 32). Describing our perception of it is just another way of describing how we can navigate it. Presence is access, perception indicates action, and to all of phenomenological experience, whether visual, tactile, auditory, or otherwise, 'there corresponds a different repertoire of skills for exploring and so achieving access to what there is' (Noë 2012: 33). The world is accessible or present to us in a variety of ways, and that variety is correlated with the variety of sensorimotor relations we can hold to it.

Perception is not in the perceiver, but in the world itself, in the different ways by which the world can be navigated. This is Bergson's account of 'pure perception'. Bergson calls it 'pure' because it exists only in principle, abstractly. Concretely, perception is always mediated and completed by memory, which is to say that the present – what actionists call the world as available – is both enabled by and laden with the past. I return to this idea later.

Even what we might otherwise consider to be strictly passive or receptive physiological possibilities – like the fact that the human visual field, determined by the machinery of the optical apparatus, is comprised out of only a very narrow range of the light spectrum – are also testament to and completed by possible actions (MM 68). This idea has two parts. First, sensory qualities, the data of what we might call passive perception, are contractions of enduring phenomena measured by the degree of 'tension' in a given consciousness or perceiver. These contractions generate both continuity and quality.

Perceptual continuity is an effect of what empirical researchers call a 'flicker fusion rate' (see Hecht and Smith 1936 and Veitch and McColl 1995). Flicker fusion occurs when separate perceptual instances no longer appear distinctly, in succession, but run together continuously. This rate marks the limit of perceptual contraction, designating the number of repeated instants the system in question is capable of apprehending distinctly. For human beings, flicker fusion occurs at 60 Hz, which means that our visual apparatuses are capable of registering just under sixty elements in a perceptual second. Electronic displays with refresh-rates of more than sixty times a second appear smoothly, while displays that refresh or flicker at lower frequencies appear discontinuous. Unified perceptual instances are effects of the physiology of the visual apparatus. They are actions performed on matter. The same effect of fusion and the generation of perceptual continuity can be measured in auditory sensations between cycles of eighteen and twenty per second (Békésy 1960: 258–9); and, again, in tactile impressions as well, though rates vary relative to where the impressions are localised (Allen and Hollenberg 1924).

In addition to perceptual continuity, contraction also produces sensory qualities out of quantities (MM 25). The perception of colour is a paradigm case. It is the effect of the contraction of the (4 trillion/ second) oscillations of electromagnetic light waves of different lengths (or frequencies) into a sensible field, running across a spectrum from red to violet (see Wyszecki and Stiles 1982). We perceive colours by 'condensing enormous periods of an infinitely diluted existence', the electromagnetic waves themselves, into separable perceptual impressions (MM 275). 'Every concrete perception, however short we suppose it, is already a synthesis . . . of an infinity of "pure perceptions" which succeed each other' (MM 238). Hypothetical beings whose perceptual fusion rates flicker across time scales radically slower than ours might be able to watch all of human history summed up in a moment; and beings of higher rates, discerning too many oscillations of light in each impression to have them condensed in colour, would live out perceptual lives of pure quantity, registering the vibrations and repetitions beneath what we call sensible matter and nothing more (CM 157–8). The same point obtains, just as it did for the effect of continuity, across our other sensory capacities as well (MM 268–9; cf. Prelorentzos 2008). Auditory qualities, the differentiation of sound waves into tones and timbres, olfactory sensations, and tactile impressions are all products of their fusion, the contraction of quantity into quality, or time qualia (see O'Callaghan 2013; Latour 2004; Allen and Weinberg 1925; and Hirai 2018: 8).

The passive physiology of contraction responsible for the generation of perceptual quality and continuity indexes a set of actions performed on matter, but they are actions actually performed, executed in the production of quality and continuity, not virtual in their perceptual presentation as other possibilities to act are. Perception always means virtual action. Sensory qualities, actions already performed, bear upon possible actions by furnishing their possibility space with a material substrate and distributing across it the qualified objects in which a body's other capacities are instantiated. The two operations are logically but not temporally distinct. We perceive at one instant a qualified object as well as what we can do with it. Its qualities are generated by the way our perceptual systems act on matter, and the object is present for us as a set of actions that can be performed.

There is an evolutionary-ecological explanation for why our perception-images are qualified in some ways and not others. Qualities are contracted from quantitative repetitions – like the oscillation of light waves – in accordance with the degree of tension that characterises consciousness, determined through the frequency fusion rates of a sensory system and their coordination in action. That degree of tension is also the index of the 'freedom' or 'intensity' of a form of life, a measure of how much it can do in excess over an otherwise automatic reaction to stimulation (see Ansell-Pearson 2005a: 1113; cf. Panero 2012). It is 'the primordial function of perception' to 'grasp a series of elementary changes under the form of a quality or of a simple state by a work of condensation. The greater the power of acting bestowed upon an animal species, the more numerous, probably, are the elementary changes that its faculty of perceiving concentrates into one of its instants' (CE 301).

Concentration means contraction. To contract external time (like the vibrations in a light wave) in a perceptual experience (of colour) is to distinguish one's own duration from the slower vibrations of matter. This temporal asymmetry introduces a gap between the subject and the world, between mind and matter or subject and object (MM 279). The time-scale gap is what conditions the qualities of experience. Matter loses discrete quantity in order to produce new qualities through temporal fusion. We see colours and hear sounds as an effect of the tension or time scale at which we live our lives in time.

The relation between the physiological generation of perceptual images and the ability to act on them is dramatised in *Creative Evolution*'s critique of adaptationist explanations of the emergence of complex organs, like the eye. Bergson observes there that light is sometimes spoken of as if it had called the eye into being on its own. The suggestion is that an environmental fact, the mere presence of light, might suffice to explain the existence of the perceptual organs that are receptive to it. This idea suggests a correspondence between the perceptual apparatus and its images, but at the same time disarticulates the apparatus from its basis in sensorimotor ability.

Receptivity to light, at least in locomotive animals, means being able to navigate a light-saturated environment. 'Our eye makes use of light', as Bergson puts the point, 'in that it enables us to utilize, by movements of reaction, the objects we see to be advantageous, and to avoid those which we see to be injurious' (CE 71). That requires a prolongation of the retina 'in an optic nerve, which, again, is continued by cerebral centres connected with motor mechanisms' (CE 71). It makes less sense to suppose that light also caused the formation of the other systems – nervous, muscular, osseous – that are 'continuous with the apparatus of vision in vertebrate animals' (CE 71). What looks like passive physiological receptivity, the sensitivity of a part of a body to some part of the world, is indissociable from the way that body is able to navigate that world, acting upon it. Action and perception are co-implicated, from behaviour programmes and learned skills all the way down to the perceptual apparatuses themselves.

c. Affordances

The idea that the body's capacities for action virtually configure its perceptual world corresponds to what the ecological psychology literature calls abilities and affordances (see Gibson 1979: 127–46; cf. Posteraro 2014). The action possibilities that show up in an environment for some organism are called affordances. 'The *affordances* of the environment are', in J. J. Gibson's terms, 'what it *offers* the animal, what it *provides* or *furnishes*, either for good or ill' (1979: 127). Pens afford bodies that can properly manipulate them the ability to write. Lamps afford bodies endowed with the right perceptual systems the ability to navigate environments absent natural light. Ability and affordance are inextricably linked, defining each other, a fact that 'implies the complementarity of the animal and the environment' (1979: 127). The climbability of a staircase is an affordance only for animals that can climb it (Cesari et al. 2003: 111–24). The ability to climb determines the climbability-affordance of environmental features. The same features might afford other things to other animals, but if they cannot climb, then the world cannot show up as climbable for them.

Affordances multiply in tandem with an increase in freedom, in capacities to act. The more affordances an organism can perceive in a material object, the less dependent it is upon that object's actual properties, having instantiated in them a larger virtual set. Affordances are virtual actions. In interacting with its environment, the organism does not merely recognise a set of affordances – as if they were there already, awaiting it – but instantiates those affordances in its environment virtually and then actualises them in the process of performing its abilities. Experience happens between the two, actualising an affordance as it executes an ability (see Noë 2004: 216).

A body's abilities also configure its world in physical form, through the dynamics of niche construction. An organism's niche, according to one philosopher of science, is 'the set of situations in which one or more of its abilities can be exercised' (Chemero 2009: 148). In *Triple Helix*, Lewontin defines a niche as 'the penumbra of external conditions that are relevant to [a given organism] because it has effective interactions with those aspects of the outer world' (2000: 8–49). Developmentally, animals evolve selective sensitivities, in tune with their sensorimotor abilities, to relevant environmental features in a niche. The animal's niche will influence the development of the animal's sensorimotor abilities in turn. The animal's activities alter its niche, and its niche influences the development of its abilities. While the organism unfolds itself into its environment, its environment *enfolds* itself back into the organism. This is Niche Construction Theory (see Levins and Lewontin 1985: 88–9 and Odling-Smee et al. 2003).

Niche Construction theorists Post and Palkovacs (2009) provide an example of this dynamic in their observation of the way differential patterns of predation led to different populations in two populations of guppies. That population differential meant different patterns of excretion, and those patterns produced fluctuations in algae growth that modified selection on guppy colour through an increase in carotenoids derived from the algae, in a process that biologists call 'evolutionary feedback'. While a change in colour alone does not imply a corresponding change in ability, it is easy enough to conceive of a case that does. Earthworms are known to alter the makeup of the soil in which they live. A change in soil can mean a difference in selection pressure on local plants. Different plants mean different niches, and a difference in the kind of sensorimotor abilities local organisms are able to manifest. In the very manifestation of those abilities, the layout of an organism's environment is changed, both for itself (the earthworm) as well as for others (species of plant).

On the developmental scale, this reciprocity is manifested in the operation of natural selection on the genetic composition of a population of organisms relative to its environment. Selection pressures from the environment alter the genetic composition of the population. Changes in the population rebound on the environment in turn. 'Organisms and environments', in Lewontin's words, 'are both causes and effects in a coevolutionary process' (2000: 126). As genetic composition changes, so do the actions of which organisms are capable; and as those capacities change, so do the environmental niches in which they are situated.

The way these rates of change are linked, the way genetic composition and environmental layout influence each other over time, is itself temporally dynamic. In order to interact with a feature of its environment, the perceptual rhythms or frequency fusion rates of a given organism must sync with the duration of that feature. Meaningful environmental features are temporally relative to the organisms that interact with them. A mountain can afford climbability only on sufficiently short time scales. Over geological time, its contours become viscous. The body, its capacities, and the perceptual world articulated around it fluctuate in tune with rates of change in environments, species, and their coevolutionary interaction. Perceptual selection opens onto fields of dynamic ecological relation.

Implication

a. Contraction

Behavioural dispositions, action schemata, and operations of contraction – whether perceptual, psychological, or ontological – involve the second facet of virtuality, implication. There are some instances in which a whole can be present in one or each of its parts. In those instances, the mereological relation is one of virtual implication. The whole is virtual in its parts by being implicated in them. There are three such instances in *Matter and Memory*: the presentation of the whole material aggregate behind any actual perception of it (MM 28, 30, 276–8); the containment of an action schema, or a system of behaviours, in each of its individual movements (MM 112, 138–9); and the whole of the past implicit, or implicated, in any present action or perception (MM 191, 215, 220).

There are precedents in medieval philosophy for this virtual mereology. Boethius's category of 'potestative' wholes is one (see Arlig 2011: §2.1–2.3). John Duns Scotus's work is another that has already received some attention in connection with Deleuze. Scotus conceived the virtual as a potency-based differentiating principle for immaterial forms, such as mind, angels, and God (see Duns Scotus 2000: 334–7). Bergson's relationship to medieval philosophy remains understudied, but it has been noticed. Michael Grosso claims that Bergson's theory of duration was anticipated by Boethius (2015: 90). Milič Čapek also suggests that Bergson owes a debt to Boethius regarding the relations between eternity, inexhaustibility, and completion (1971: 385–6). Čapek holds that already in Boethius there exists the idea of an open-ended whole. Richard Cohen traces the motivation for Bergson's theory of intuition back to medieval debates around the proper relation between reason, revelation, intellect, and will (1999: 25). In her account of Bergson's position with respect to the history of philosophy, Chimisso gestures briefly in the direction of Bergson's debt to medieval thought, but does not develop the influence (2008: 52).

Boethius's category of potestative wholes is distinct from the integral and universal wholes of medieval philosophy. An integral whole involves a collection of separable parts, while a universal whole indicates an abstract classification or general category under which other individuals could fall, like species under a genus. Potestative wholes are defined by powers or functions that are neither composed out of nor divisible into proper parts. The parts that make up this form of whole are inseparable from it, but at the same distinguished by the different powers they display or actions they can perform as parts. The potestative whole is implicated in its parts without being reducible to them, and those parts instantiate the whole without being fully separable from it. Bergson's virtual involves a mereological theory of just this sort.

The first appearance of mereological implication in *Matter and Memory* refers to the way the material world insists behind particular perceptual representations of it (see Miquel 2012). In the act of perception, the perceiving subject seizes upon something that outruns perceptual images. The objects of perception are real, but diminished; their sensible qualities are effects of their contraction, an operation by which a more dilute existence is condensed in a unified and qualified perceptual experience. Present behind both the narrowed object and its qualities are the complete material aggregate and the quantitative repetitions, or vibrations, that are contracted in its qualities – not actually, but virtually. Perceptions are parts of a larger whole, and the whole from which perception dissociates its parts is virtually given along with them.

The whole is virtual in its parts just as possible actions are virtual in the objects upon which they can be performed. This virtual mereology allows Bergson to reconcile his realist philosophy of perception with the claim that perception is always selective relative to the practical interests and capacities of a given body. First, because those practical interests and capacities are defined by a certain frequency fusion rate, a principle for the contraction and qualification of matter; and second, because it is nonetheless possible for an interested body to overcome the perceptual limitations that it has imposed on the material world.

Perceptual selections are kinds of actions performed on the material world. They are performed actually, not virtually instantiated in their objects as possibilities. Virtuality inheres here not at the level of parts, as it does for possible actions, but at the level of the whole – behind, beneath, or beyond the perceptual images that are its parts. If this were not the case, then how could we ever come to discern more in a material object over time? How could we ever draw more of the world into perceptual presence? There must be more there, obscured (for now) by our selective interests, but nascent and capable of being brought to actuality by the expansion of an action space, a sophistication in learned ability, a refinement in perceptual acuity, its technological enhancement, and the scientific coordination of techniques and technologies with each other. If our perceptions are parts of the material whole, and if the difference between being and being perceived is a difference in complexity, but not in kind, then the whole is there in its perceptual parts (MM 28, 30). It is implicated there virtually, capable of being realised through modifications in the perceptual filters that particularise it.

Bergson says the same about qualities. Behind the perception of one colour the other colours near it on the spectrum are virtually suggested (CM 157). Contractions retain their genetic impresses, which means that we can imagine how they might be re-dilated and opened onto the larger fields that they contract. Differences in colour are the effects of 'the narrow duration into which are contracted the billions of vibrations which they execute in one of our moments' (MM 268). If we could stretch those durations out, live them at a slower rhythm, then we should be able to watch a given colour 'pale and lengthen into successive impressions, still coloured, no doubt, but nearer and nearer to coincidence with pure vibrations' (MM 268–9). At a slow enough rhythm, we would register nothing but successive vibrations. That imaginary operation testifies to the virtual implication of the vibrational continuity of the colour spectrum in any one of its perceptual determinations. The perception is, on the one hand, in me, the perceiver, as the contraction of a vastly diluted phenomenon in a single moment of my duration; but, on the other hand,

> if you abolish my consciousness, the material universe subsists exactly as it was; only, since you have removed that particular rhythm of duration which was the condition of my action upon things, . . . sensible qualities, without vanishing, are spread and diluted in an incomparably more divided duration. (MM 276)

Virtual implication is a matter of contraction. The whole is contracted in its parts; it insists there virtually, implicated in the actuality of each of them.

b. Themes

Behaviour programmes affirm the same mereological principle. To recognise an object is to know what to do with it, how to act on it. To act is to play out some set of those possibilities in and through perception. The perception of a recognised object always prolongs itself into movement. It *involves* a 'sketch' or diagram of the movements through which the possible actions reflected in it can be actualised, that is, *evolved* (MM 111). That 'sketch' is what we perceive in objects that we recognise. Perceptual recognition consists in a conditioning of the present by the past. If perceptions are prolonged into movements, and if movement diagrams are the result of an accumulated past consolidated into motor tendencies, then, practically, what we perceive when we recognise is not the present but the past.

Perception is of the past. It acts itself out in the present. Together they comprise the duration of experience. That means that there is no such thing as 'an' experience, something present and identifiable that you could have, undergo, register, retain, or recall. Experience is a quality of the accumulated past. Experience conditions the present, suffusing perception with virtual possibility and association, enabling increasingly individual action.

Experience is not made up of an atomistic series of present perceptions. The same thing is true of the possible actions that infuse perception with the past. Possible actions come in sets or programmes – that is, wholes. These wholes are virtually implicated in each of their parts; 'prefigured' in each movement through which they are played out, the way a melody resides in each of its notes (MM 112). This kind of virtual prefiguration is an achievement. It has to be learned, accumulated, and consolidated. We do that by repeating a series of actions until they become habitual. Then it can be said that the whole series is virtually implicit in any one of its parts. The whole is there, 'imprinted' on memory, 'stored up in a mechanism which is set in motion as a whole by an initial impulse' (MM 89, 90).

Every perception has its organised motor accompaniment. These are 'consolidated' in the organism, linking sense stimuli through the intermediary of the nervous system with a set of corresponding movements (MM 112). Perceptions do not correspond one-to-one with the actions that would complete them, but are structured through diagrams or organised sets of actions. Take the ordinary matter of walking through a door. One perceives all the relevant information almost immediately. As soon as the door is recognised, all the right tendencies are there, nascent, pressing into actuality. Hesitate before performing them and a feeling of discomfort arises as the perception–action circuit has been interrupted. The body wants to act out the right schema. It is there, virtually implicit not only in the object of recognition, but in itself as well, 'preformed' in each of the movements through which one acts it out, enters the room, removes one's coat, and so on (MM 112).

To recognise an object is to contract into a simple perceptual impression a whole suite of behaviour programmes, and to acquire or 'have' a behaviour programme is to contract into any one of its steps, phases, or movements the whole of which it is a part. Contraction instantiates the more in the less – again, just as a whole melody is 'in' its notes. Virtual wholes are like themes. They give sense to their parts, organising them while remaining irreducible to any of them taken separately. To learn a skill, memorise a poem, or acquire an ability is to unify a series of otherwise discontinuous movements under a theme.

The opposite is true as well. In the process of learning a physical exercise, we can begin by watching it performed and proceed by trying to reconstruct it. But while the visual perception was of a continuous whole, the movements by which we try to reconstruct it are compound, made up of a series of muscular contractions. The movements through which we try to copy the image are 'already its virtual decomposition' (MM 137). That means that the continuous whole of visual perception 'bears within itself, so to speak, its own analysis' (MM 137).

The same point obtains for learning how to discern words in the auditory impression of a foreign language, and of learning how to speak that language. Virtual in confused perceptions are the elements into which we decompose them in order to build them back up intelligently. In so doing, particularly in cases when we are trying to learn how to discern or perform another's actions, we proceed by structuring our confused impressions with the right motor diagrams, the systems of behaviours that organise and make intelligible what the other person is doing.

At the same time, overlaid on the motor mechanisms contracted through repetition are memories, more or less vivid, of each lesson, each attempt at memorisation, each phase in the process of learning. These memories are not motor; they are not a matter of habit, and they do not form a continuous whole to be contracted in each one of its phases. Recall a first swimming lesson; now another; now another. Each is unique, dated, individual; 'I can see it again with the circumstances which attended it then and still form its setting' (MM 89). The swimmer can act out the strokes she learned in her lessons, one after another, just by introducing her body into a pool of water; but her memory of an individual lesson 'is a representation, and only a representation' (MM 91). It can be filtered down to some set of details marking the event, like the smell of the public pool in which she was taught, or the frustration she felt in failing to replicate the instructor's movements. Or it can be enlarged to the point of encapsulating the whole day in a single picture. The habit furnished by that lesson requires a certain amount of time to develop, as each movement is acted out in the right order; but the representation of the lesson itself can be assigned any duration whatever, and shortened or lengthened at will. While we speak of an imprint on memory in both cases, the habit is 'lived and acted', more properly considered 'part of my present', while the representation or recollection of a particular phase in its acquisition is 'like an event in my life; its essence is to bear a date, and consequently to be unable to occur again' (MM 91, 90). The representation has an affinity with the past itself, apart from its utility for action.

c. Memory

Of the two forms of memory, it is the second that Bergson considers to be true memory, that is, 'memory *par excellence*' (MM 95). The first, that of the motor diagram, is '*habit interpreted by memory* rather than memory itself' (MM 95). Between the two there is a difference in kind. Every image is doubled as it occurs (see Worms 1997: 102–3). As a perception-image, it is consolidated in the motor mechanisms of the body, deepening the body's habits, adding to its familiarity with the objects around it, and extending its virtual actions. At the same time, the image is added to the past from which it can be recalled as a memory. 'Either the present leaves no trace in memory', as Bergson explains, 'or it is twofold at every moment, its very up-rush being in two jets exactly symmetrical, one of which falls back towards the past whilst the other springs forward towards the future' (ME 99). This doubling operates through a directional differentiation. Difference in direction defines a difference in kind (MM 102; cf. Lawlor 2003: 32). The perception-image is directed towards action, through the prolongation of perception in the behavioural schemata that organise and complete it; the memory-image, on the contrary, is directed towards the past, being preserved in memory.

Habits inhibit recollections, filtering out from the past what is not useful or relevant to the sensorimotor present. At the upper limit of recognition, where habits complete the objects of perception, recollection does not take place consciously at all and there is no intervention of memory-images. And yet, 'our past psychical life is there', virtual (MM 113). Even when the past is inhibited by the continuous circuit of perceptions and actions, 'this memory merely awaits the occurrence of a rift between the actual impression and its corresponding movement to slip in its images' (MM 113). It does that in most cases to help reinstitute the circuit between action and perception, feeding perception the images it needs to activate the right behavioural schema, as in cases when one needs to pause to recall the details of some particular lesson in order, for instance, to continue playing a game. In these cases, memories are recalled based either on a resemblance they share with perception-images or because of their utility for some current system of actions. The present is still primary over the past, and memories are recollected on its basis, actualised in a series of stages 'by which this image [of the past] gradually obtains from the body useful actions or useful attitudes' (MM 168). While it looks on the one hand as if the present is reaching backward into the past to find there the image it needs to complete its prolongation into action, at the same time 'the virtual image evolves [from the past] towards the virtual sensation, and the virtual sensation towards real movement' (MM 168). The present takes from the past, and the past exceeds the ways by which it makes itself useful to the present.

To remember the past is to actualise it in an image in the present (see Lawlor 2003: 36, 53–8). Yet there is more to the past than what can be recalled from it in the present. The past exists in its own right as well, as 'pure memory', which is Bergson's term for the existence of the past independently of what we actually remember of it (MM 163,

176, 181). If the past exists in its own right, then it grows larger with every passing present that is added to it. If the present is perceptual and oriented towards possible actions in the future, then it is really nothing but the 'invisible progress of the past gnawing into the future', swelling as it advances (MM 194).

Referring back to *Matter and Memory*, Bergson writes in *Creative Evolution* that just 'as the past grows without ceasing, so also there is no limit to its preservation' (CE 4). It is not limited by the constraints of embodied (or en-brained) memory, the subjective ability to recall and retain. Rather, 'the piling up of the past upon the past goes on without relaxation [and] is preserved by itself, automatically' (CE 4). The past is preserved in its entirety whether we remember it or not – real without being actual: virtual.

Bergson defines the present in terms of a sensorimotor – or practical and embodied – attention to life, sustained by an effort to 'make present' that which interests us (CM 127; cf. Lapoujade 2008). The present is as variable as the field of attention with which it is coextensive, corresponding to the effort that generates and fixes it. The present involves an operation of division. While the effort to divide can be relaxed or intensified – varied – its limits are circumscribed by the degree of durational tension typifying its subject (see Pöppel 1978). That degree of tension defines the interest an organism is capable of taking in its environment – just as was the case for the selection of perception-images from matter. In this case, the limit of interest or use marks the threshold between present and past. It can be attenuated through the modulation of interest or attention.

That the past is past just means that it is no longer of sufficient interest to remain present. But it can come rushing back involuntarily when the 'attention to life' determining the present is upset, suspended, or unnecessary – as in intoxicatory rapture, reverie, dream states, or near-death experiences (CM 127). The involuntary influx of the past back into the present suggests that the past requires no special faculty to secure its preservation. It is there, dormant, while we attend to present purpose. Even while in a state of non-actuality, the past is not in a state of *unreality* as well, and does not therefore require an entity from which it could derive its being – like the brain's hippocampus or frontal cortex, actual structures in which it would reside in the form of images.[1]

Here is the Bergsonian inversion. If it seems that the past is abolished in its passing, it is because 'we shut our eyes to the indivisibility of change, to the fact that our most distant past adheres to our present and constitutes with it a single and identical uninterrupted change', and on that basis 'we think ourselves obliged to conjure up an apparatus whose function would be to record the parts of the past capable of reappearing in our consciousness' (CM 128). The opposite is true. The brain does not store up the past; it turns us away from it, recalling only simplified portions of anterior experience in the form of memory-images in order to complete present perception. The brain is a mechanism for the 'canalisation' of attention in the direction of the future. The brain's function is to choose from the past, to simplify it, attenuate, diminish, and utilise it, but not to preserve it (see Čapek 2016). The brain is an operator of division. The brain allows us to cut into the stream of time that is our duration, dividing away a past by rendering it dormant and so focusing the organism on its present circumstances. Time is continuous. Its division is effectuated through a modal differentiation performed by attention (see Burton 2008: 328–39).

Bergson employs two images for the relation of the present to the past. The first is the set of concentric circuits for attentive recognition (MM 128). The second is the famous inverted cone whose tip is inserted into a plane (MM 211). Attention divides the present *from* the past, but redoubles the present *with* the past as well, 'by reflecting upon it either its own image or some other memory-image of the same kind' (MM 123). The present's 'own image' is the memory-image that doubles it, generated just as it is perceived, streaming off in the direction of the past just as the perception-image directs the body towards the future. Bergson calls this a 'virtual image', sometimes equating it with the 'afterimage' of a perception or the memory of the present, and at other times with the phenomenon of déjà vu or false recognition (MM 168, 124; ME 99–108). Each formulation refers to the idea that memory is produced in the present, not after it as a faded residue of perception (as Hume and the empiricist tradition believed). At every instant there is a doubling or scission in images, between what will have to be recalled from the (virtual) past and what is to be perceived in the (actual) present (see Deleuze 1989: 68, 79). Déjà vu, the feeling that one has already experienced what one is currently experiencing, is not a misfiring in recollection but a disclosure of memory in its true operation, generating images contemporaneous with the present images that they double. Attention leverages this doubling in the direction of an intensified perception of the present.

Bergson rejects the classical model of attention that imagines the mind made up of a quantity of light to be concentrated and focused on some particular object instead of diffused across a wider array of them (MM 123; see Pashler 1998: 35–100 and Moran et al. 2016). He invokes the modal distinction between actual and virtual and locates attention between them, as the mechanism for the introduction of the one (memory) into the other (perception). As each virtual image peels itself away from the present and joins the past, the past as a whole receives and is reconfigured by it. Attention is the process through which that past is brought into circuit with a present perception to fill it out, an operation that 'may go on indefinitely; – memory strengthening and enriching perception, which, in its turn becoming wider, draws into itself a growing number of complementary recollections' (MM 123). Perceptions are narrow, or simple, and representable by tighter circles on the concentric model the less they draw upon memories in order to supplement and enhance them. The two circles expand together, away from the automatic prolongation of a perception into the action that completes and neutralises it. To attend to an object requires an interruption of the motor response mechanisms (see Lucas et al. 2012). It demands a suspension of attention to life – to the present – in order to pull an increasingly wider sphere of the past into actuality.

Attention is not concentration. We do not discern the depths of or details in an object without projecting onto it 'an actively created image, identical with, or similar to, the object on which it comes to mould itself' (MM 124). This happens so regularly that it can be difficult to distinguish memory from perception in the recognition of everyday objects, as they are filled out with memory-images. We make inferences from the past in order to anticipate the completion of the perceptual process and employ learned heuristics to attenuate the complexity of the perceptual field. As a result, the perception of an object O, on the concentric model, is rounded out by the memory-image A that overlies it, and in this way 'our distinct perception is really comparable to a closed circle in which the perception-image, going towards the mind, and the memory-image, launched into space, career the one behind the other' (MM 126).

Since each memory-image contains the whole past contracted in it, implicated virtual-mereologically, attention can pass from the memory-image immediately overlaid on its object (A) to higher circuits of memory – B, C, and D on the concentric model. The whole past is there in each circuit, and is 'capable, by reason of its elasticity, of expanding more and more', pulling into circuit with perception a greater number of memory-images and so intensifying the present with the virtuality of the past (MM 128). The expansion of memory is the unfolding of its virtual mereology, the dissociation of implicated images from out of their consolidation in an initial perception (MM 130). That dissociation, represented by an ascent in concentric circuits into the past, discloses deeper strata of the perceived object – B', C', D' – and expands the field of the perceptual present as well.

Just as the higher circuits of memory are contracted in the virtual image overlaid on perception, so are the 'causes of growing depth' in the perceptual object 'virtually given with the object itself' (MM 128). Bergson refers to these deeper dimensions as 'virtual objects' (MM 167). Virtual memories remember virtual objects, and actualised objects are deepened through their virtual supplements, which can be instantiated through recollection. Here virtuality designates the reality of the material field simplified and selected away from in the perception-image. Attention intensifies that image by electrifying it with wider circuits of the past and opening it back onto the deeper dimensions of its field.

Memory does not consist in a closed set of individual memories, but in a differentiated, dynamic field of potentials for recollection. The field of memory is planar, sheeted, or levelled (see Deleuze 2006: 61–5; cf. Lundy 2018a: 69–96). The levels of the cone are distinguished from each other by detail, affective tone, colour, texture, timbre, and rhythm. Each level repeats the memories of every other at a different rate, intensity, or speed. Higher levels allow potentially actualisable memory-images the full sweep of their detail; lower levels contract them in simplified form, diminishing personal markers and consolidating more of them at lower levels and quicker rhythms. Lower levels take less time to recall; they insert themselves into perceptions more easily, smoothly, facilitating action instead of interrupting it. Higher levels take more time to recall; at the limit, they involve reverie, hypnogogia, psychedelia. They interrupt the temporality of action and loosen life into a kind of dream. But it is the same psychical life pulsing at each level, 'repeated an endless number of times on the different storeys of memory' (MM 129). Attention is the mechanism by which the perceptual present is lived at a higher storey of psychical life, at a different degree of tension, the consequence of which is the increase in perceptual perspicacity that marks its work.

Attention is a matter of utility or interest. It brings the past into the present for a purpose, while 'the memory-image itself, if it remained pure memory [that is, past], would be ineffectual', that is, entirely virtual (MM 163). Memory-images can only become actual through their insertion into a perceptual present. To call them virtual is to call them dormant apart from the perceptions that attract them (MM 181). Their dormancy is an independence from interest or utility. The details of one's childhood bedroom are there in memory, but virtual, because they are not currently useful. Even when they are employed to complete or enhance a perception-image in the present, they still remain attached to the past by their roots, retaining something of their original virtuality even in being actualised (MM 171). Otherwise they would not be known as memories. Attention accesses the past by actualising it in an intensified present, but that present is only the most contracted point of the whole past, the point at which memories are lived in a sensorimotor present and prolonged into the movements that explicate them.

The past is analogous to the total field of material interaction. Recollection individuates the past in images just as perception individuates the material whole. The difference is that the past–present relation is bidirectional, running not only from the whole of the past to its contraction in actualised memory-images, but from memory-images back towards the whole past in turn. The ongoing virtualisation of the present through the generation of memory-images that double the images of perception implies that the past is in a state of dynamic alteration, reconfigured by every memory-image that is added to it. The actual present rebounds on the virtual past. This feedback dynamic is the third facet of virtuality. It is represented in the cone.

The cone appears first in a rudimentary form (MM 197), and then again with slightly more detail (MM 211; see Lawlor 2003: 46–7). The simpler cone represents the totality of memory (points SAB) narrowed towards its tip in the image of the body (S), which is inserted into the plane of images representing the material world (P). At the base of the cone (AB), memories are spread out at a maximum degree of dispersion, unconscious and virtual. The second, more detailed cone pictures the various planes of memory that cross-sect the descent from the base to the tip, indicating 'a thousand repetitions of our psychical life' in increasingly contracted forms and higher degrees of tension, 'figured by as many sections A'B', A'B', etc.' (MM 211). At the cone's tip is the present, the most contracted state of memory at which the past is consolidated in the body's sensorimotor apparatuses, their capacities, and the behavioural themes that enlist them. These possibilities for actions are instantiated in the plane of material images, filtering them through the tip of the cone and doubling them with the virtual memories leveraged by attention and revealed through déjà vu. Perception's virtualised memory-images are sent back through the cone in a process that simultaneously reconfigures each of the cone's planes or levels of contraction and culminates in the retroactive transfiguration of the cone's tip, the body, and its perceptual registration of images in the present.

The cone's philosophical import resides in the inversion of ordinary temporal priority. The cone offers an image of non-chronological time according to which the past coexists with every present that passes into it, along with every degree or level of its own contraction. It is not the past that succeeds the present, following from it as it seems from the perspective of present experience, but rather the present that is a dimension of the past, an effect or product of a maximum or 'infinite' state of its contraction.

Deleuze calls these the 'paradoxes' of the past (2006: 61; cf. 1989: 99). Bergson's institution of a difference between actual and virtual is supposed to resolve the paradoxes: the present is the whole past actualised, which means that the past in general coexists *virtually* with the present in which it is contracted, and *virtually* with itself in all of its states of contraction, its sheets, circuits, planes, levels, or storeys. The paradoxicality of coexistence requires the postulation of two things in the same time and place *and in the same ontological register*. Virtual coexistence subverts the second condition.

To say that the past in general – the whole cone, points SAB – coexists virtually with each of its own sheets or planes – cross-sections A'B', A'B' – is to say, borrowing a subtle terminological distinction from Deleuze's *Difference and Repetition* (1994: 207), that the virtual is differen*t*iated in itself, independently of its differen*c*iation from the actual present. The past is differentiated across each of its degrees of intensity, or levels of contraction; and it is further differenciated from its maximum or infinite contraction in the actual present. The virtual differentiation of the past in itself means that the differences in 'tone' or 'colour' that distinguish different sheets of memory are not effectuated merely by the projection of subjective preferences in the present – say, to recall a harsh experience more warmly or fuzzily, with less detail – but are rather proper to, and distinctive of, those sheets themselves, marking their degrees of intensity off from each other and defining their various levels of contraction.

Recollection, or the actualisation of memory in images, requires a 'leap' or relocation into the past, onto one of its planes and into one of its regions; it requires the transposition of the subject into memory as much as it involves the actualisation of memories in the subject (MM 173). Bergson does not himself employ the term 'leap' to describe this act; he says instead that we have to 'detach' ourselves from the present in order to 'replace ourselves' in the past – first within the past in general, and then from there, through the work of 'adjustment' (like the focusing of a camera), into a certain region of the past, onto one of its sheets (MM 171).[2]

The term 'leap' is useful for its emphasis on discontinuity, since the difference between present and past requires one to break with the former in order to enter into the latter. There is a form of contraction proper to each (see Khandker 2014: 26). The past is ontologically contracted in itself, across all the planes defining it; and the present is a psychological contraction of the whole in a present image complex. Between the two are, as Deleuze explains, 'all the circles of the past constituting so many stretched or shrunk *regions*, *strata*, and *sheets*: each region with its own characteristics, its "tones," its "aspects," its "singularities," its "shining points" and its "dominant" themes' (1989: 99). Each of the past's circles is punctuated by what can be potentially recalled from it, the potential recollections that pulse at their right rhythm on it alone, as they are dispersed across wider regions and cramped in narrower ones (being too general or too particularised). Those dominant recollections colour the memories associated with them, shining across them. The resultant dynamic qualifies the whole plane with a certain tone or feel – like the gauzy nostalgia and shimmering hues of childhood experience. Each memory constellates around the singular moments that define its proper circle and borrows from them an affective coloration, a depth and rhythm. The ontological contraction of the past in itself consists in these constellations. They are all contracted together – and explicated – in the actuality of the psychological present (see Jankélévitch 2015: 93).

Explication is not a unidirectional operation. Psychological contraction feeds back into its ontological conditions. The actualisation of the whole past in a set of images makes the past dynamic in turn (see Trabichet 2012). Perception is doubled by sets of virtual images that are sent streaming in the direction of the past. The process of recollection must be split in just the same way. Memory-images actualised in the present have their virtual doubles too. They must be sent back into the past just as well. Every event of recollection, in contracting the past in an actual image whose virtual double is introduced back into it, requires that the whole past be reconfigured to accommodate it – and at every degree of intensity. The very plane that is actualised in a memory-image is transfigured *in the past* by its own actualisation in the present.

There are some analogues for this idea in the neuroscientific research on recollection (see Wimber et al. 2015; Voss et al. 2017; Voss and Bridge 2014, 2015). The research suggests that to recall a memory is to update its contents through the terms of its recollection in the present, 'overwriting it' and so 'restoring' a basically new memory in its place. Memories left unrecalled and so preserved in their original detail are increasingly unlikely to be recalled at all. It is not hard to see how Bergson might accommodate these ideas. Memories are recalled for their usefulness to the present, and the logic of utility tends to be self-reinforcing as we rely on certain elements of the past to help us in the present. There is a melancholy here. Either our memories are recalled and updated, and we lose their original content, or else they are preserved in an oblivion from which we are less likely to save them. Despite the affinity, the two takes on memory remain fundamentally incompatible (see McNamara 1996). The past, on Bergson's account, is preserved independently of its recollection; it is not 'stored' in the brain, and its reconfiguration is an ontological event, not a neurological process.

In the spatialised language of the cone, memory-images are generated at the cone's tip in the present, and sent cascading up towards its base, rippling through the various planes they will pass through on the way. The tip re-contracts the transfigured cone in turn, so that the perceptual present virtually implicates the whole past just as the past is modified by the present in which it is implicated. If pure memory is entirely virtual, it is not for that reason causally closed to its actualisations. It is only through them that it is kept open. This is a consequence of the mereology of the past. Memory-images generated in the present and added to the past reshape the whole because the whole is contracted in each of its parts. Every part implicates the whole, which means that the addition of new parts to the whole has to reconfigure it in turn, in order to contract it anew.

The doubling and virtualisation of images in the present makes the present part of the past, intertwining the two, which means, in Alia Al-Saji's words, that 'the past as a whole reverberates with every virtual image and is reorganized as a result' (2004: 218). Far from a static repository of images past, like a container into which present experiences are deposited or a tape on which they are recorded, 'the "past in general" consists of dynamic and transformative planes' in open processes of modulation, fluctuating according to alterations in their degrees of tension and relaxation (2004: 218).

There are therefore two moments in the philosophy of the memory cone. First is the thesis that the actualisation of images consists in the contraction of the whole past in a sensorimotor present (MM 173, 319, 322). This is the descendent ontological model of actualisation: from the past to the present, or virtual to actual. The second moment consists in the disruption of the virtual closure of the past through an inverse causal reaction of the virtualisation of the present back onto the past, reorganising it as a whole. In this moment is the dynamism, or open-endedness, of the mereology of the virtual. The actual is produced as a contraction of the virtual in a movement that reacts on the organisation of the virtual in turn.

Dynamics

a. Invention

In 'Intellectual Effort', an essay from 1902, Bergson completes *Matter and Memory*'s discussion of the actualisation of memory-images and extends that theory in the direction of what, five years later, will become the *élan vital*. Examining the earlier study of effort helps to clarify the bidirectional relationship between the virtual and the actual. I call this virtuality's dynamism. It is at work in instantiation and implication as well.

'Intellectual Effort' is a study of psychological contraction. The preliminary datum of the study is the 'dynamic scheme', which Bergson also refers to as a schematic, abstract, or simple idea (ME 123, 126, 131).[3] All the elements that are to become determinate images exist in the scheme dynamically, or pre-individually, 'in the estate of reciprocal implication'. Determinate images actualise the scheme by evolving it 'into parts external to one another' (ME 124). The question is how that operation is to be effectuated. Bergson locates its operation in the feeling of effort that marks terse, concentrated thinking.

Bergson begins with memory and concludes with the effort of invention. He examines two cases of the former. From the first, a study of the skilful chess-player, he extracts the 'dynamic schema' and establishes the mereology of effort on its basis (that is, the way in which effort dissociates determinate images from their implication in a unified whole). In the second, the experience of trying to recall a forgotten name (or trying to follow an ill-understood language), he locates the feeling of a 'direction' in which the schema is to be developed in order to arrive at its determination in images. Invention presents a domain in which actual images can be seen to turn back on their dynamic schemas, modifying and opening them to novel development.

The real difference between an amateur and a master, according to Binet's 1894 study of the psychology of chess, is whether one can play multiple games simultaneously, while blindfolded.[4] The blindfolded player has the moves of her opponents indicated to her; she plays a piece in response, moving from one game to the next, while apparently keeping in mind the distribution of pieces on each board. If the amateur finds this feat impossible, it is because her skill remains dependent on the concrete actuality of a particular game and the representation of each position on its board. The master can abstract away from these particularities and into a realm in which she does not require visual representation, a realm in which the calculations through which she carries out her moves take place as if independently from each of her opponents. As Siegbert Tarrasch expressed the point: 'some part of every chess game is played blindfolded. For example, any combination of five or more moves is carried out in one's head – the only difference being that one is sitting in front of the chessboard. The sight of the chessman frequently upsets one's calculations' (qtd. in Hearst and Knott 2009: 183). Blindfolded chess is chess purified, so far as possible, of an ultimately unnecessary visual stimulus.

Before Binet, the best explanation for the ability to play chess blind seemed to belong to Taine, who argued that the blindfolded player employs a heightened visual memory, maintaining and updating a representation of each game as it is played. 'Evidently', he wrote, 'the figure of the whole chess-board, with the different pieces in order, presents itself to the players at each move, as in an internal mirror, for without this they would be unable to foresee the probable consequences of their adversary's and their own moves' (Taine 1871: 38).

Bergson sees in Binet's response – according to which the player does not maintain a visual representation of each game but instead has to make 'an effort of reconstruction' at every move – the suggestion that what is held in mind is not a set of distinct images but rather their interpenetration in an abstract whole (ME 123).

The abstraction does not consist in the determinate qualities or visual details of each piece, but of what they can do, and in what way their powers interact, conflict, and compose. The game is grasped as a set of relations 'between allied or hostile powers' (ME 123). On the basis of an abstract grasp of these powers, the blinded player is supposed to be able to visualise each element of the whole, the pieces and their positions on the board, without having to maintain the image of any one individually. This whole is what Bergson calls a dynamic schema. It is as complete as any of the images through which it is actualised, but it does not contain those images in itself; it rather suggests them, implicates and implies them.

The feeling of this kind of effort is supposed to consist in the push to dissociate distinct parts from out of their implication in the virtual unity of the whole. It has as its essence 'the evolving of a scheme, if not simple at least concentrated, into an image with distinct elements more or less independent of one another' (ME 125). It is a vertical or descendent movement: from the scheme towards the images implicated in it. From an analysis of the struggle to recall a forgotten name, Bergson concludes that we would never be able to accomplish such a feat were it not for the indication of a direction in which effort is to be expended to arrive at the determination of the image, like an impulse common across a set of distinct elements. It is this 'directive' quality to the scheme that allows it to unify a multiplicity of elements without effacing their distinctions entirely (ME 140). It unifies by traversing and organising its elements, like a melody across its notes. It fuses them, but nevertheless indicates how they are to be dissociated, that is, in what direction their dissociation is to be pursued. While the scheme is simple (or unified), and therefore distinct from the images to be elaborated out of it, it still *tends* into that elaboration, suggesting the images without for that reason resembling them.

Bergson sees in the experience of invention a final aspect of the psychology of effort (ME 131). To invent – following Ribot's account in *L'Imagination créatice* – is to solve a problem creatively, without supposing a determinate solution in advance. Effort in invention consists in the work through which the gap between the problem and a possible solution is filled in, bringing a real solution about. The solution, if it is to be the object of invention, can be known at first only schematically; and so 'invention consists precisely in converting the scheme into image' (ME 132).

What is special about inventive effort is the fact that it is opened onto an uncertain horizon in view of which the inventor experiments within a range of potentialities disclosed (as images) by the invention's dynamic schema. This is how we are to understand Bergson's term, the 'unforeseen', as a necessary epistemic concomitant of novelty (ME 133; cf. CM 8). The novelty of invention resides in the retroaction of images on the schema from out of which they are drawn, 'the movement by which the image turns around toward the scheme in order to modify or transform it' (ME 133). The inventive scheme is altered through the process of its dissociation into the images that represent its actual products. Those images are tested in the course of the inventive process, postulated provisionally, experimentally modulated, combined, revoked. But they 'are always reacting on the idea or the feeling which they are intending to express' (ME 133). The movement is bidirectional as the two evolve in tandem: changes in the scheme imply differences in its images, and changes in those images retroact on the scheme they are meant to realise. The scheme contains dynamically what is accomplished, and therefore static, in its images; and in invention the scheme is worked over again, 'modified by the very images by which it endeavours to be filled in' (ME 132).

Bergson draws from this study the claim that psychological effort elides the binary causal alternative between compulsion and attraction, revealing each – at least in the intellectual register – to be an extreme limit of a more fundamental intermediate activity. The causality operative in this form of activity is best understood mereologically. The movement between the scheme and its images is a movement between the qualitative multiplicity of the whole and the determinate actuality of its parts. It consists 'in the gradual passage from the less realized to the more realized, from the intensive to the extensive, from a reciprocal implication of parts to their juxtaposition' (ME 143).

Processes of actualisation – the externalisation of parts from out of a state of their mutual implication – retroact on the virtual whole actualised through them. Traversing the planes of memory means reconfiguring the planes themselves. Converting a scheme into images involves the alteration of both in turn. Causal orders run in each case from the virtual through the actual and back. This bidirectional dynamism is a general feature of virtuality, operative in each of its dimensions. In the remaining sections of this chapter, I examine the dynamic nature of instantiation and implication in turn.

b. Affordances

Possible actions are instantiated virtually as affordances in the perceptual field, acting as a filter on its complexity. The actualisation of those possibilities reconfigures the network of affordances as a result, reacting on their own virtual layout. The dynamic relationship between the virtual and the actual, abilities-affordances and actions, fluctuates over the time scales of development and behaviour.

Anthony Chemero, a proponent of the kind of enactivism anticipated by Bergson's theory of perception, argues that a given organism's sensorimotor abilities – Bergson's virtual actions – become sensitive to relevant features in the organism's environment over time, adjusting to and syncing with them (2009: 150–1). The virtual instantiation of possibilities as affordances in the perceptual field takes place developmentally. Even once provisionally accomplished, it remains variable over time. Possibilities for action are developed, and the virtual conversion of those possibilities into environmental affordances is attendant on the duration of that development. On the shorter time scale of behaviour, our 'sensorimotor abilities manifest themselves in embodied action that causes changes in the layout of available affordances' (2009: 151). Actualised possibilities retroact on their virtual status as affordances. The causal relationship runs in both directions, since changes in a layout of affordances imply correlative changes in which and how possible actions can be executed.

It is easy to see how actual acts could modify actual features of the environments in which they are performed. It is harder to see how the unexercised capacity to act can be causally influenced by a change in the layout of virtual affordances or vice versa. As John Protevi observes, 'only the act of climbing a tree, not the unactualized ability to climb, can knock some bark off the tree or strain a muscle' (2013: 151). True enough. But changes in any one element of an ability-affordance field imply corresponding changes in the others, both virtually and in their actual counterparts. These changes and interactions take place first, it is true, on the level of the actual. The animal's ability to climb only impacts the layout of its environment when exercised. 'It is', in Protevi's words, 'only these individuated actions that can change the [virtual] web of relations' in which their possibility is instantiated (2013: 152). Their actualisation resolves a field of related elements in some one determinate activity or event. In so doing, they make future determinations of the same sort more likely than others. They reconfigure the lines of differentiation available, modulating the insistence of other available potentialities, carving out for each a kind of depth in a topological possibility space canalised in favour of some actions over others.

Given that abilities correspond to virtual affordances – and that the manifestation of affordances necessitates the presence of abilities – changes in the actual, whether somatic or environmental, also correspond to reorientations in the layout of the virtual affordances they resolve. Changes to these fields – ability-affordance complexes – manifest as changes both to action capacities as well as to the environmental affordances that complement them. When an animal's activities impact on the physical layout of an environment (as when a species alters its niche over evolutionary time), those changes reorganise the virtual field individuated in that activity. A different virtual field means a corresponding change in actual activity.

If the developmental relation between abilities and affordances concerns the way organisms shape and are shaped by their environments in the activity of living, on a behavioural scale an organism's activities alter the organism's perceptual world as well (Chemero 2009: 152). Given some set of functional concerns (say, in swimming), a perceptual world or practical field is composed of those features salient for the ongoing navigation of some environment (a body of water). A swimmer has to coordinate her kinaesthetic activity with changes in her practical field, synchronising the rate at which what she can do changes – for how long breath can be held, how many strokes can be performed before a rest is necessary, and so on – with changes in environmental affordances (strength of current, temperature of water).

Behavioural and developmental time scales are reciprocally related. Developmental causality is constituted by the ability-affordance relations that occur on the behavioural time scales of individual organisms. Developmental changes constrain the way organisms behave practically. Abilities and affordances causally relate in two temporal iterations, though they are co-defining and inextricable. This bidirectionality testifies in both to the dynamic interaction by which the actualisation of the virtual reconfigures the virtual for future actualisations.

c. Performances

The actualisation of a behavioural theme, or action scheme, interacts with the virtual whole implicated in each of its movements in a similar fashion. Performance theorists suggest that there is a reciprocal relationship between themes – whether musical, dramatic, ritual, or otherwise

– and the variations through which different individuals and collectives enact them (see Schechner 2013: 30). Performances are events. They do not take place 'in' some other medium – the way novels take place in words and on paper or screens, or paintings take place in paint and on a canvas – but between bodies in concert with artifacts and technologies, in shifting situations and relational contexts. There can be no performance of a theme that is not already necessarily a variation upon it, however slight. There is no performance of a theme that is not also an implicit reference to its last iteration as well, as the theme does not so much exist 'behind' its performances as much as it runs through each of them, accumulating the variations through which it is performed in each instance.

One music theory scholar argues that variations do not vary themes per se; 'what they vary are reductions of themes' (Cook 2016: 204). Themes run through each of the moments that would otherwise count as variations upon them. The postulation of the theme behind its variations requires a reduction of that theme down from the structural level where its variations are located. Even admitting the institution of a structural difference between the theme and its variations, it does not follow that the relation between the two is unidirectional, as if the theme were just a reduction down from its variations or its variations were just a complexification of the theme varied, whether analytically (in the mind of the critic) or in their performed musical differences. Between each structural level of the performance is the 'hint of a reciprocal determination' by which one influences the other while being simultaneously influenced by it. Performances are performances of other performances as much as they are performances of themes.

To act is to actualise a behavioural theme or action scheme one movement at a time. The theme organises those movements virtually, giving them their sense and direction. Mereologically, the theme, as a qualitative whole, is implicated in each of the movements through which it is performed, as its parts. If the performance of a theme always involves variation, then the whole's parts are in a state of constant even if minimal flux. If the whole is implicated in each of its parts, if it is what organises them, then the whole must be reconfigured in tandem with the performative modification of its parts. Performances, as variations on the theme that they perform and that informs them, are reincorporated into that theme. Themes are actualised through the same movements that react upon them, made dynamic via the causal relay of their own actualisations.

Were it not the case that actualisations reconfigure the themes they actualise, then it would be difficult to explain how we improve our abilities, modulate our habits, or integrate our behavioural tendencies. We seem to do so by reincorporating differences in the movements through which a behavioural theme is actualised back into the theme itself, preparing it for further differentiation in some direction through subsequent actualisations. In acquiring a skill, one learns to discern new features in the shifting relations between one's body, relevant tools, other actors, and the environment, bringing new virtual instantiations to bear within the perceptual field. As one becomes appreciative of those differences through the institution of a new behavioural theme, one learns to simplify them down into actionable stimuli-response circuits, integrating what are at first dissociated perceptions and responses into organised diagrams for action. That organisation happens in the actual, through the process of reconfiguring our bodily dispositions and rearticulating our perceptual fields in view of some goal or exemplar and in response to some set of problems. Achievements made in the actual are then reincorporated into the virtual theme traversing them. That theme impels further differentiations to be made in the direction of competence and then mastery in actual behaviour.

In the case of learning, the bidirectional interplay between theme and movement, or part and whole, ratchets upward in integration and complexity. In the case of the degradation of a learned skill, or the loss of a set of habits, the causal circuit tends towards disintegration or the dissociation of actions from perceptions. In both cases there is a dynamic relation between the theme and its performance, the whole and the parts in which it is implicated, the virtual and the actual.

Mereological implication is dynamic in the same way as the instantiation of possible actions in the objects of perception is dynamic, and again as the contraction of the whole past in every passing present. In each instance the virtual whole is kept open by its actualisations, the processes through which its parts are externalised in order to be reincorporated with a difference.

Conclusion

The idea of virtuality runs through a number of interrelated issues from *Matter and Memory* and into 'Intellectual Effort'. In each case the virtual refers to a part–whole relationship. The material whole insists behind its articulation into perception-images, which are dissociated as parts from out of the whole by passing through the filter of perception, whose screen is meshed by the body's virtual capacities for action. Those capacities are implicated as parts in unified behavioural wholes of their own. The whole of the past insists behind the memory-images of it as well, whose parts are dissociated from out of the whole of memory for purposes of utility. Their logic is defined by a sensorimotor attention to life (whose own nature is defined by the first sense of virtuality). The virtual whole is in dynamic relation with its actualised parts.

The virtual also defines what is nascent in each case, what tends towards actuality while structuring and exceeding it. The virtual frames the actual. It filters perception through the actions that complete it, organises those actions into thematically unified sets, and introduces the images attendant on those sets into the whole of the past, reconfiguring the past and transforming each of its determinations in turn. The instantiation of possibilities in perceptual objects are given as affordances. As they are actualised through their execution, they are reconfigured as well. This happens both behaviourally and developmentally. The implication of a behavioural whole in the parts through which it is performed is reconfigured in the movements that actualise it. Performances always involve variations on the theme that organises them. Those variations are reincorporated into the theme, modulating it in turn. The whole past is transformed by, just as it is causally responsive to, its actualisations in the present. The whole is in each case always open, in dynamic relation with the parts into which it is dissociated, in which it is contracted, and in response to which it becomes.

This is the theory of virtuality that I attribute to Bergson and employ throughout this book. In the next chapter, I distinguish the virtual from the alternative modal categories of actuality, possibility, and potentiality, and use virtuality as an interpretive frame for making sense of Bergson's idea of tendency. Tendency is one of the central concepts of *Creative Evolution*, though Bergson works it out across each of his texts. I show why it should be understood in terms of his virtual mereology. The last three chapters of the book put that idea to work in laying out the rest of Bergson's philosophy of biology.

3

A Discourse on Tendency

The central concept in my reconstruction of Bergson's philosophy of biology is the idea of tendency. Though Bergson develops this idea throughout his work, he does not provide it with systematic exposition. In this chapter I trace the idea of tendency across Bergson's major texts and interpret it according to the virtual mereology of Chapter 2. Then I parse the idea of tendency according to recent work in the metaphysics of powers. I hold that tendency is directional inclination, the essence of movement. Its modal status is neither actual, nor possible, nor potential, but virtual. Tendencies are qualitative multiplicities, virtual wholes tending into concrete actuality through processes of bifurcation, dissociation, and the determination of parts.

Introduction

This chapter develops the metaphysics of tendency in Bergson's philosophy of biology by interpreting it through the modal mereology of the virtual. The rest of the book builds out his philosophy of evolution on this basis. Chapter 4 puts the metaphysics of tendency to work in a series of reconsiderations of Bergson's views on the time/space and life/matter dyads. In Chapter 5, I show how the idea of tendency is what lies behind Bergson's image of the *élan vital*.

I begin by situating the idea of tendency against the background of Bergson's study of psychological effort. The concept of tendency is meant to capture a form of effort de-subjectivised, dislocated from personal psychology and generalised across all forms of directedness. The chapter traces the development of the idea of tendency across Bergson's work, from its initial employment in *Time and Free Will*'s introspective account of free acts through to its formulation as a concept in its own right, in Bergson's last texts. Bergson works out the features of tendency first according to the structure of time and the dynamic reality of movement. In *Creative Evolution*, he provides tendency with a metaphysical extension, making it intrinsic to becoming in general. In its mature formulation, tendency is directional inclination. It is what accounts for the open-ended development of the evolutionary process.

By setting tendency apart from more familiar modal categories such as possibility and potentiality, I show how tendency accounts for actualisation processes while preserving the ineliminable novelty of duration. Bergson's theory of tendency involves the mereology of the virtual. Tendencies are qualitative multiplicities, virtual wholes tending into concrete actuality through processes of bifurcation, dissociation, and the determination of parts. In this regard, the actualisation of a tendency is more like the determination of the past in the form of particular memory-images than it is like the manifestation of a disposition or expression of a power. The actualisation of a tendency involves a process of dynamic contraction. This distinguishes the Bergsonian view from other theories of tendency. I conclude the chapter by considering one recent example in the metaphysics of powers at some length. I argue that Bergson's conception avoids some issues that persist through the contemporary account.

Intellectual effort and élan vital

Anyone who knows Bergson even superficially will know something about the *élan vital*. While any relatively attentive reading of *Creative Evolution* should yield the position that the *élan* is not a vital principle of the traditional variety, headway nonetheless remains to be made in how to understand it. I return to the issue of vitalism in Chapter 5. My focus for now is on Bergson's designation of the *élan* as an image, and as an image for tendency in particular.

Whatever else can be said about *élan vital*, it is clear that Bergson understands it to be an efficacious feature in the organisation and evolution of organisms. It is also clear that the *élan* is supposed to elude the binary alternative between efficient and final causality – or mechanism and finalism – and so cannot be determined in the causal terms recognisable to either (see Sinclair 2020: 213–15). Bergson suggests that it is better thought through what he calls an 'image' instead, and of an impetus or an '*élan*' in particular (CE 257). He explains why in a letter to Floris Delattre.

> The image intervenes most often because it is indispensable, none of the other existing concepts being able to *express* the author's thought; the author is then obliged to present it *suggestively*. . . . To give just one example: when I relate the phenomena of life and of evolution to an 'élan vital', it is in no way an ornament

of style. It is even less meant to mask in images our ignorance of
the deepest causes, as when the vitalist in general invokes a 'vital
principle'. . . . The truth is that philosophy only offers philosophers two principles of explanation in this matter: mechanism and
finalism. . . . Now . . . the place to be is somewhere in between
these two concepts. How should we determine that place? I have
to point to it, to indicate it since no concept between mechanism
and finality exists. The image of an *élan* is only this indication.
(Bergson and Delattre 1936: 399)[1]

I spent some time discussing Bergson's idea of image in the last chapter.
There are several kinds of image – perception, memory, and now indication or index. The last kind is used as a tool for the expression and
communication of thoughts that frustrate or escape the delimited boundaries of available concepts. Images can be used to point to something
outside of concepts, prompting thought to think beyond habituated
forms. Yet, as tools for the expression and communication of thought,
images are also necessarily breaks or circumscriptions of thinking; and so
with them 'you can no more reconstitute thinking than with positions
you can make movement' (ME 55). While the need for an image arises
when thinking outstrips the concepts that are available to it, any particular image, as a satisfaction of that need, is only produced 'when thinking,
instead of continuing its own train, makes a pause or is reflected back on
itself' (ME 55). The image is something like an unstable intermediary
between the intuition of something beyond the domain of concepts and
the concepts beyond which it is directed.

The *élan* works as an image by pointing beyond the already-existing
concepts through which biological phenomena can be thought. It is
not a concept with an identifiable object to which it could correspond.
When Bergson describes the operation of the *élan vital* as the driving
force of the evolutionary process he is not speaking literally, as if to
postulate the existence of such a force. He is invoking an image, indicating that there is a certain x at work in driving and patterning evolution
while being irreducible both to the efficient causality or impulsion of a
mechanism as well as to the final causality or attraction of a pre-existent
goal or aim.

Though there may well be a number of possible sources for this
particular image both within and outside of Bergson's thinking, I am
partial to the idea that he drew the image of the *élan* forward into *Creative
Evolution* – and into its formulation as the *élan vital* – from the psychology
of effort, whose details he elaborated five years earlier in 1902's 'Intellectual
Effort' (see Posteraro 2021b).[2] On this reading, the *élan* is an image for the

operation of life borrowed from the psychological domain because 'no image borrowed from the physical world can give more nearly the idea of it' (CE 257).

Recall that according to Bergson's psychology of effort, the feeling of effort consists in the psychological act of straining to dissociate distinct parts, or images, from out of their implication in the unity of a dynamic scheme. The activity of dissociation is supposed to be guided by the indication of a direction, a suggestion of how the images implicated in the scheme are to be developed. This 'directive' quality is what allows the scheme to unify a multiplicity of elements without effacing their distinctions. The operation of effort is at its most acute in the act of invention that suspends an already given relation between the scheme and its images. Effort in invention consists in the transfiguration of the scheme through the development of its images, just as those images are determined on the scheme's basis in turn. The two evolve in tandem. Changes in the scheme imply differences in its actual images, and changes in those images retroact on the scheme they are meant to realise. The movement is bidirectional. Its products are effects of reciprocal adjustments in both of its poles.

Bergson argued that psychological effort is something other than compulsion or attraction. These two forms of causality are – at least in the psychological register – the opposite and extreme limits of a more fundamental intermediate activity. The causality operative in this form of activity should be understood mereologically and multi-modally. The movement between the scheme and its images is a movement between the qualitative, virtual multiplicity of the whole and the determinate actuality of its parts. It consists 'in the gradual passage from the less realized to the more realized, from the intensive to the extensive, from a reciprocal implication of parts to their juxtaposition' (ME 230).

Bergson concluded 'Intellectual Effort' by indicating that the operation of psychological effort is exemplary of 'the very operation of life' (ME 230). This is what, five years later, he will call the psychological interpretation of life in *Creative Evolution*. It is the idea that life can be modelled psychologically, that the *élan vital*, the impetus behind evolution, can be imagined as a kind of 'psychological cause' (CE 91). Psychology is the appropriate domain in which to make sense of life because 'the essence of the psychical is to enfold a confused plurality of interpenetrating terms'. This is something that psychic phenomena share in common with life, and something that marks both off from the possibility of a fully physical determination. Bergson has interpenetration in mind when he says that 'life is of the psychological order' (CE 257).

The psychological image also points to something beyond psychology. It indicates a form of effort – vital effort – that is non-subjective, detached from the creative will of an individual and relocated at the level of evolution as a continuously unfolding process. Bergson attempted to capture something of this form of effort in the creation of a concept as well. 'Life', he says, 'is tendency' (CE 104). It has been said recently that tendency is 'the most essential concept that Bergson advances in order to understand something of the impulse' of life (Sinclair 2020: 215). The concept of tendency is supposed to comprehend that for which the *élan vital* is an image. Tendency is like effort but de-subjectivised. It is to life what effort is to mind. The *élan vital* is an image that points to it. The rest of this chapter develops the idea of tendency. I return to the *élan vital* as part of Bergson's 'psychological interpretation of life' in Chapter 5.

The development of an idea

The idea of tendency recurs across Bergson's work (see Vollet 2012: 362–73). He first introduced the term in *Time and Free Will* (TFW 30, 104, 132, 138, 209). It was initially motivated by the issue of how to conceive free acts against the alternative between determinism and indeterminism. According to that binary, all actions are either determined by the causes that preceded them, or they are free only insofar as they are spontaneous exceptions to the natural order and therefore causally inexplicable. Bergson argues that agential acts are better conceived in terms of expression instead of causation or spontaneity (TFW 165–6).

Actions express inner multiplicities of psychic states. Each state, abstracted from the inner multiplicity and considered on its own, can be identified according to a tendency towards certain outward results, such as the expression of anger through violence (TFW 132). But inner life is not directed by disparate and self-identical states whose outcomes in action are given in advance. Each psychic state blends into the others, 'borrowing an indefinable colour from its surroundings' (TFW 132). It is impossible to predict how these states could be externalised in actions. It is up to action to decide that, by dissociating tendencies from their mutual implication and expressing one of them in determinate form. Every action expresses tendencies, but the tendencies themselves do not dictate their expression in action beforehand. Actions should be thought free to the extent that they express a whole personality, manifesting its deepest tendencies and flowing from out of its most fundamental aspects (TFW 167). Actions should be thought unfree to the extent that they express only some one overriding tendency or a superficial and passing set of them.

Freedom is best conceived in terms of degrees, on a spectrum from profound to superficial. This idea is best appreciated from the perspective of the tendencies that actions express. Actions come to be defined by the dominant tendency that characterises them. This tendency appears, after it is expressed in action, as if it were the cause of the action that expresses it. The action appears to be an effect of antecedent conditions. The causal attribution operates retrospectively, on the basis of the tendencies already embodied in some determinate configuration in the action that expresses them.

Although it is left somewhat underdeveloped, there is a concept of tendency at work behind these ideas. Tendency refers first to heterogeneous multiplicities made up of interpenetrating psychic states. For such a multiplicity to tend is for it to lean, push up against a limit, move in a direction. These multiplicities tend in a variable set of directions, changing and developing as they endure. The actions that express them dissociate individual tendencies from the psychic whole in which they interpenetrate. Only after the fact can they be re-attributed to the psychic whole as distinct things in numerical combination (TFW 132–3). They are fused in an original state, and individuated via the act of externalising them (TFW 104). Tendency also refers to the individuated impulse that is embodied and expressed in a given action. In this second sense, tendencies are closer to catalysts than causes, and closer to potentials than possibilities. Once dissociated from the psychic whole, tendencies to act might even be identified with potentials – so long as they are not defined by an already given end in advance.

Matter and Memory provides the psychological conception of tendency with a physiological extension (see Worms 2000: 28, 30, 33–4, 61). The first reference to tendency in *Matter and Memory* consists in the idea that perception-images are always accompanied by the motor tendencies, or movement diagrams, that bring perception into circuit with action (MM xiii). Perceptions actuate a set of action possibilities. The performance of some subset of those actions completes the perceptual activity that prepared the body for them. Perception has 'its true and final explanation in the tendency of the body to movement' (MM 41). Tendency means physiological disposition, readiness for or availability to the actualisation of some set of action possibilities. It appears to be defined by automaticity, as movement or motor tendencies are 'triggered' by the perceptions whose possibilities they realise (see Sinclair 2020: 92–6). But just as with the psychological conception, motor tendencies cannot be specified by already given ends. They are like a body's push into a set of contingent outcomes. Tendency is directedness towards action in general, without being oriented in the direction of any particular action in advance.

The automaticity of tendency characterises intellectual habits too. The intellect is defined by the tendency to spatialise, decomposing movements and changes into stable things, states, and properties (MM 161). The perception of movement triggers spatialising cognition. This is an 'invincible tendency' (MM 154). That means it is a default state, always operative. To overcome it, the intellect 'must do itself violence' by 'revers[ing] the normal direction of the workings of thought' (CM 160). Spatialisation is the 'normal direction' of intellectual activity. It is derived from the 'fundamental tendency' of all living beings to extract from their situations whatever is most useful for them (MM 154, 158–60, 218). According to Bergson's naturalised epistemology, cognition is adapted to the practical dictates of how to act on matter, not the theoretical task of passively registering the truth (CM 71; cf. Madelrieux and Schott 2020). Spatialising makes the objects of cognition actionable. This evolved, utilitarian orientation is what grounds the spatialising tendencies of the intellect and the physiological dispositions, or bodily schemata, that structure perceptual experience as well.

In *Time and Free Will*, tendency has a conceptual affinity with freedom. Actions are free to the extent that one's deepest psychological tendencies are expressed in them. In *Matter and Memory*, tendency is allied with automatic movement, motor mechanism, bodily disposition, and habit instead. The two might seem like different uses of the same word, but there is an inner link between them. In both texts, tendency is oriented towards action in general without being teleologically specifiable in terms of any particular actions in advance. The idea of action consists in the introduction of novelty (see Lawlor 2018; cf. Riquier 2010). Bergson clarifies this conception of freedom in *L'évolution du problème de la liberté*. 'Freedom is creation', he says, in the same sense that 'there is creation in a work of art'.

> This does not mean that the work of art is production *ex nihilo*, in that it is made of nothing. By [creation] we mean that, given the conditions which contributed to the production of the work, the work, if it is truly the work of an artist, adds to these conditions something absolutely new and absolutely unpredictable. (PL 7th Lecture)

To act in this sense is to perform in a way that cannot be predicted as a necessary causal outcome of past events or conditioning stimuli. In the terms of *Matter and Memory*, action involves the ability to hesitate between an array of possibilities. Action is free according to how much it can detach itself from its causal conditions, or how wide a sphere of possibilities it can choose within. The tendency to movement that defines the body is what brings perception into circuit with these action possibilities. The spatialising tendencies that define the intellect facilitate choice by making dynamic phenomena cognisable as objects to be acted upon. The freedom that is at stake in *Time and Free Will* is a freedom of acts carried out by the body in space. The freedom of the free act is a measure of the psychological tendencies embodied in it, but these tendencies are tendencies to movement or action in general all the same.

After *Matter and Memory*, Bergson begins to employ the idea of tendency with a more focused intention. The idea appears twice before *Creative Evolution*. The first is in a lecture course on Leibniz's *De rerum originatione radicali* delivered in 1898 (see Vollet 2008: 286–92). Leibniz's ideas about tendency were likely mediated for Bergson by the work of Maine de Biran and Ravaisson (see Dunham 2016, Sinclair 2019: 70–9, Devarieux 2012). In line with the French tradition of interpretation, Bergson attributes to Leibniz the aim of explaining not only what is, or what exists, but how the things that exist could have been possible in the first place (L 37). Leibniz is supposed to have uncovered beneath the existence of actual things not only their logical possibility, but a dynamic inclination, a *conatus* or striving to exist (L 42). This inclination towards actuality is supposed to constitute a non-actual ground or 'root' beneath the existence of actual things (L 52). The originating root of things is their possibility.

There may be two forms of possibility in Leibniz. In addition to the inert, preformist conception that Bergson criticised under the category of finalism, there may be something of a generative theory in Leibniz as well. Possibility has a 'quantity of reality' or 'quantity of essence' of its own (Leibniz 1989: 487). The reality of possibility is its tendency to become actual, a view referred to as Leibniz's Doctrine of Striving Possibles (see Shields 1986). Leibniz also distinguished between passive capacity and active force, which he called tendency (see Kistler and Gnassounou 2007: 17–18). Active force is not yet actuality but what will become actual if there is nothing to stop it. He claimed that his idea of 'active force differs from the mere power familiar to the Schools', Aristotle's potential, since this latter form of power 'is nothing but a close possibility of acting, which needs an external excitation or a stimulus, as it were, to be transformed into action'. By contrast, active force contains its own conatus, an inclination towards action. 'It is thus carried into action by itself and needs no help but only the removal of an impediment' (Leibniz 1969: 433). This idea of a power that is 'carried into action by itself' if it is not interfered with, frustrated by conflicting powers, or constrained by the medium for its expression is strikingly close to the view of tendency that *Creative Evolution* will develop.

Bergson seemed to appreciate the idea that beneath the generation of actual things, at the root of the world, there is tendency or inclination at work (see Vollet 2012: 367). Perhaps he dislocated this metaphysical feature of the idea of tendency from the other facets of Leibniz's thought and reconfigured it in his own terms. Perhaps he was inspired to extend his own conception of tendency in a metaphysical direction, encoding it in the structure of reality instead of just the mind or body – as he had done in *Time and Free Will* and *Matter and Memory*. Whatever the nature of the historical influence, two things are clear. On the one hand, Bergson was interested by Leibniz's idea of a tendency to actuality. But on the other, his commitment to an ultimately creative reality ruled out any such idea in the form that Leibniz had given it. If beneath the actuality of a thing is its tendency to exist, that tendency cannot resemble the thing in advance. Neither can it be pre-specified in view of the end of realising that thing, like a seed and the tree it will become. Tendency should admit of no such predictability. This means that it cannot constitute the essence of the thing of which it is the tendency. The actualisation of some thing does not complete, resolve, or exhaust the tendency that it actualises. Actuality, for its part, is not static or inert either. Tendencies should be considered inherent throughout the actual as well. Actualisation is creative, an introduction of novelty into the world.

Five years later, in the 'Introduction to Metaphysics', Bergson began to develop an idea that satisfied those requirements. He employs an explicitly metaphysical conception of tendency for the first time in his work. It has many of the same features accorded to it in *Time and Free Will* and *Matter and Memory*, but now it is attributed to reality in general. 'This reality', Bergson writes, 'is mobility' (CM 158). All mobility or movement is dynamic with tendency (CM 131).

> There do not exist *things* made, but only things in the making, not *states* that remain fixed, but only states in process of change. Rest is never anything but apparent, or rather, relative. The consciousness we have of our own person in its continual flowing, introduces us to the interior of a reality on whose model we must imagine the others. *All reality is, therefore, tendency, if we agree to call tendency a nascent change of direction.* (CM 158–9)

What Bergson said of inner life in *Time and Free Will* he now ascribes to reality in general. The ceaseless, interpenetrating flow of lived duration is the conscious index of an ontological fact. Reality is movement. Stability and thing-hood are epiphenomenal effects (CM 122). Bergson agrees with Leibniz that the 'non-actual ground' or 'root' of reality is tendency. Tendency is the inner truth not of a set of possibles that resemble their actual forms in advance, but of the movement and change that define a world in the making, a world of absolute mobility and only relative stability. 'When I speak of an absolute movement, it means that I attribute to the mobile an inner being' (CM 131). The inner being of movement is not defined by a fixed direction that its movement would trace out, as if by realising a pre-existent diagram. The tendency that defines movement is itself variable, or *directional without being directed*. In Bergson's formula, tendency is a 'nascent change of direction' (CM 159). It is actualised across the particular directions that an unfolding reality takes. As there is no end to mobility, there is no point at which tendencies cease to be effective in and through the making of actual things.

The theory of tendency

By 1903, Bergson had developed most of the elements required for a complete theory of tendency. *Creative Evolution* is the first text in which that theory is deployed in something like a systematic fashion. Bergson begins by bringing together his previous uses of the term. He recalls *Time and Free Will*'s conception of psychological duration as 'a flux of fleeting shades merging into each other', and adds to it *Matter and Memory*'s account of the contraction of the total past in every passing present, filtered according to the dictates of practical attention (CE 3, 5). The result is that 'our past', since it is all there virtually, 'is made manifest to us in its impulse; it is felt in the form of tendency' (CE 5). The ongoing contraction of psychological duration in conscious experience is a matter of tendency. That tendency is consolidated in our character, felt in the tendency towards movement, and concretised through the perception-action circuits of *Matter and Memory*. According to *Time and Free Will*, freedom resides in its most complete expression in action. Bergson concludes *Creative Evolution*'s opening discussion of psychological life by summarising: 'for a conscious being, to exist is to change, to change is to mature, to mature is to go on creating oneself endlessly' (CE 7). When he asks whether 'the same [should] be said of existence in general', we know from 'Introduction to Metaphysics' that the question targets the metaphysical status of tendency and that the answer has to be yes (CE 7).

This is not the first occasion on which Bergson made that suggestion in that form. 'Intellectual Effort' concluded with a similar provocation. Bergson returns to it in *Creative Evolution* and develops the point under the idea of tendency. He conceived the work of effort in the earlier text according to 'the gradual passage from the less realized to the more realized, from the intensive to the extensive, from a reciprocal implication of parts to their juxtaposition' (ME 230). This movement from 'intensive to extensive', from 'reciprocal implication' to 'juxtaposition' of parts, is supposed to characterise 'the very operation of life' (ME 230). *Creative Evolution* completes the equation: 'life', writes Bergson, 'is tendency' (CE 99). It is not one particular tendency among others, but tendency itself. The characteristics of life are the characteristics of tendency and vice versa. What Bergson said about effort, and what he says about life, should apply to his theory of tendency in turn.

Tendency is now said to have an essence: 'to develop in the form of a sheaf, creating, by its very growth, divergent directions among which its impetus is divided' (CE 99). It is only partly right to describe the essence of tendency as arborescent ramification, whose passage traces Darwin's tree of life. More fundamentally, the essence of tendency is to incline towards actuality, to tend to become actual if unhindered. What that inclination towards actuality looks like when it is allowed to take place is, of course, unpredictable. The idea that tendencies tend to differentiate in the form of a sheaf elaborates upon the idea, from 'Introduction to Metaphysics', that tendency is nascent change of direction. Nascent variability means directional differentiation, an inversion of the Leibnizian view that tendency is the inclination of a possible existent to exist in actual form. The development of a tendency runs instead from a pre-individual inclination through a series of divisions in the directions it takes. That is because, contra Leibniz, tendency is essentially temporal, tending towards an indeterminate future. 'This we observe in ourselves, in the evolution of that special tendency which we call our character' (CE 99).

Tendencies may develop in the form of Darwin's tree of life, but the essence of their development is not Darwin's descent with modification. Tendencies remain continuous through bifurcation, like a personality through the various choices that constitute it: 'a unity that is multiple and a multiplicity that is one' (CE 258). One tendency dissociates into a ramifying series of tendencies, dividing itself between them without being separate from them. Tendency is both one and many, one implicating many, realising or extending itself in myriad. The idea recalls the gradual externalisation of images from out of the qualitative multiplicity of a dynamic scheme. Like the scheme and its images, the original state of a tendency is to implicate an interpenetrating multiplicity of what will become distinct tendencies once they are dissociated. This is why Bergson sometimes speaks as if tendencies exist in the plural, as a set of divergent directional inclinations, and other times as if there is only one, tendency as nascent change of direction. Tendency, nascent change of direction, is dense with directional potentials.

Individuated tendencies, like the tendency towards complex vision in some species, are like 'psychic states, each of which, although it is itself to begin with, yet partakes of others, and so virtually includes in itself the whole personality to which it belongs' (CE 118). Even after being developed, tendencies retain a continuity with each other because they preserve and contract their history of development. Since they develop from a state of interpenetration, 'there is hardly a manifestation of life that does not contain in a rudimentary state – either latent or virtual – the essential characters of most other manifestations' (CE 112 tm). Particular tendencies realise in concrete actuality, such as in some species, elements that remain virtually continuous with each other. What elements? Bergson says that the tendency towards mobility characteristic of animal life retains, in a state of suppression, the tendency towards torpor that characterises the plant (CE 118). Mobility and torpor are made possible by developmental channels whose actual products are unpredictable. The development of one retains the unrealised potential for the other as an adult retains the personality possibilities she left unexplored in childhood. Plants that evolve the ability to move might testify to this. I return to the idea in Chapter 6, in a discussion of evolutionary convergence.

The idea of a unity that is multiple recalls *Time and Free Will*'s distinction between qualitative interpenetration and its spatialised derivatives. Bergson seems to have that distinction in mind when he claims that 'in its contact with matter, life is comparable to an impulsion or an impetus', that is, a tendency (CE 258). 'Regarded in itself it is an immensity of virtuality, a mutual encroachment of thousands and thousands of tendencies which nevertheless are "thousands and thousands" only once regarded as outside each other, that is, when spatialized' (CE 258 tm). Life is tendency, which, like inner life, is composed of an interpenetrating continuum of directional inclinations (CE 100). That continuum can be individuated, the tendencies that compose it spatialised and extended each in directions of their own, tending towards deviating limits. Then life looks like a set of many different tendencies, not one impulsion but a multiplicity.

Bergson reached a series of conclusions regarding spatialisation in *Time and Free Will*. Once spatialised – thought through the spatial frames of extensity, exteriority, and succession – the various 'states' of our inner experience appear 'side by side in such a way [that we] perceive them simultaneously, no longer in one another, but alongside one another [in] the form of a continuous line or chain, the parts of which touch without penetrating each other' (TFW 101). For this reason, inner experience is better thought on the models of an organic whole and a musical melody instead. The first because our past and present states do not appear to us as exterior to one another, but integrally related, forming a whole irreducible to any of its parts taken independently of each other. The second because as we live them these states melt into each other, each one contracting the whole without fully realising it, penetrating the others and informing them in turn (TFW 100). When it comes to inner experience, it would be more appropriate to 'conceive of succession without distinction, as a mutual penetration, a solidarity, an intimate organization of elements, each of which represents the whole, and cannot be distinguished or isolated from it except by a thought capable of abstraction' (TFW 101 tm).

Bergson's equation between inner life and life in general has received an impressive amount of attention (CE 258; cf. Tellier 2008: 428–9, Gunter 1982: 637, and Caeymaex 2013: 54). The idea is that life or tendency in general is defined by the qualities of psychological duration, abstracted away from a consciousness in which they would represent choices, character development, and the manifestation of a deep, unified self-history in a cascading variety of impulses, behaviours, and actions over time (TFW 125–6, 139; CE 99–100). The kind of causality that Bergson discovered in effort was supposed to exist somewhere between efficiency and finality, impulsion and attraction. He conceives the causal efficacy of tendency in just the same way. It is, on the one hand, a *vis a tergo*, an originating impulsion (CE 103). But it is not for this reason to be reduced to the mechanism of an efficient cause. Tendencies take directions 'without arriving at ends' (CE 102). Their causal nature is directional, immanently purposive. In this sense there is something finalistic about them (see Cunningham 1914: 649–50).

Yet this is a form of purposiveness that is not oriented in the direction of distinct purposes specifiable in advance. Tendency is irreducible to the concepts of final and efficient cause, though it involves aspects of both. Bergson's conception of tendency repeats one of the central features of his conception of time, its retention–protention structure, the idea that concrete duration always retains its past history of development while opening that history onto an unforeseeable but directed future (see Riquier 2008: 298). 'Life', as tendency, 'is a continually growing action', modelled on the effort to recall and create (implicating both the past and the future respectively) but de-subjectivised and extrapolated to the evolutionary process in general (CE 128).

Creative Evolution makes two other features of the idea of tendency explicit. The first is that tendency's directionality runs across a spectrum between two directions. Particular tendencies tend towards particular limits but cannot be identified with ends, since they tend to differentiate as they develop. The finality of a tendency is immanent to the development of that tendency, temporally open to the future. Mark Sinclair suggests that tendencies are defined by a form of 'pre-teleological directionality', directionality prior to direction, like purposiveness without purpose (2020: 216). Matthias Vollet identifies the two directions that tendency can take as movement and its inverse, inhibition (2012: 363). I consider them both forms of movement, oriented in different directions, towards immanent ends. In this respect I depart from both Sinclair's and Vollet's interpretations.

Tendency is directionality in general, but on my view there are two directions in which a tendency can tend, two ways it can move. Bergson designates each with a different series of more or less interchangeable concepts, but prefers the life/matter dyad throughout *Creative Evolution*. 'Life is a movement', 'materiality is the inverse movement' (CE 249). Life means vitality, novelty, creativity, duration, movement, individuality, ascent, intensity, interpenetration, indetermination, freedom. Materiality means undivided flux, repetition, inhibition, determinism, stasis, space, externalisation, entropic homogeneity, fixity, descent. None of these terms are ever fully realised in actuality. They are movements, directions. Evolution tends in the direction of increasing individuality and free activity. Unorganised matter tends in the direction of spatiality, tending but only tending towards a mechanical order of repetition and determinism.

The second idea is that tendencies tending towards one end of the spectrum do not merely depart from but exist in directional conflict with tendencies tending towards the other. Every concrete actuality, every material body, is made up of both movements, defined by a relation of predominance or subordination between them. All of existence is a *modus vivendi* between tendencies vital and material in conflict (CE 250). Tendencies tend to become all that they can be, realising themselves as states, properties, or things so far as they are uninhibited. Inhibition is what keeps a tendency tending. As tendencies are not yet things, they can be inhibited only by competing tendencies, or else by the strictures or parameters of their modes of expression (CE 13). It is only in relation to conflicting tendencies that a tendency can be defined at all; otherwise it would have already accomplished itself and would no longer *tend*.

This was one of Schelling's insights. Where we posit the existence of a tendency, we necessarily posit the existence of a tendency that opposes it. 'For we can think of force [or tendency] only as something finite', he says, because as infinite, unlimited, it would be realised already and a contradiction in terms. Yet a tendency cannot be 'finite by *nature* except insofar as it is limited by one opposing it' (Schelling 1988: 37 and 2000: 8–9; cf. Deleuze 2004: 35–6). All determinate individuals – as the expressions of tendencies – are composed of an internal opposition, bifurcated across two contrarieties. The principle of non-contradiction states that two contraries cannot be true of the same composite in the same way at the same time. Schelling's reformulation of this principle requires that one tendency predominate over and suppress the tendency that opposes it. Only one of a set of conflicting tendencies can be predominate in a determinate individual at a given time. For one tendency to be active or dominant is for the other to be passive or subordinate to it. It is the dominant tendency that defines the nature of a composite entity.

In the 'Final Remarks' of his last major work, *The Two Sources of Morality and Religion*, Bergson formalises the theory of tendency in the form of two immanent developmental laws. The first, 'the law of dichotomy', states that for any set of conflicting tendencies there was prior to their opposition a primitive tendency that contained them both in itself, 'as two photographic views, so to speak', of its complete nature (TS 296). The converse of this law states that every pair of conflicting tendencies has its genetic condition in the bifurcation, or 'splitting up', of an antecedent tendency (TS 296).

The derivation of new tendencies should not be understood on the basis of possibilism's pre-existent programme, as if the antecedent tendency contained within itself each of the directions it would take after splitting in the form of possibilities prior to their actualisations. If that were the case, then ancestral tendencies would have included from the outset the sum total of all possible directions taken over the course of evolutionary time. Everything would have been given in advance, and time would index only the progressive manifestation in the actual of what was pre-existent in the form of the possible. To the contrary, it is a tendency's bifurcation that creates the pair of directions subsequently taken by it (TS 294). The law of dichotomy relies on a retrospective attribution. Prior to its bifurcation, there is only one unified tendency in operation. After it is divided into a set of two, it can be said retrospectively to have contained them both within itself, in a state of fusion. Properly formulated, the law of dichotomy states that all sets of conflicting tendencies 'embody two views, now rendered possible, of a primordial, indivisible activity, which was neither the one nor the other, but which becomes retroactively, through them, both at once' (TS 294).

The second law, 'the law of twofold frenzy', states that each tendency in a conflicting set, once dissociated from the tendency that contained them both, is determined by 'the imperative demand . . . to be pursued to the very end', as far as possible (TS 296; cf. Miquel 2007a: 52). The second law is supposed to explain the first. It is in the essence of tendency to extend itself as far as it can. Beyond that limit, 'it divides instead of growing' (CE 99). This is why tendencies divide, in order to continue developing, along different lines, past a threshold that marks the fullest possibility of one of them. It is the same impetus, or directional effort, that is sustained along each direction. Tendency wins the ability to extend itself further at the cost of having to split itself up (TS 297).

In *Creative Evolution*, Bergson formulated as an initial law a variation on the same idea. 'When a tendency splits up in the course of its development, each of the special tendencies which thus arise tries to preserve and develop everything in the primitive tendency that is not incompatible with the work for which it is specialized' (CE 119).

Tendencies do not unfold linearly, across a unitary, progressively complexifying line of ascent. They are diffused across a web of divergent trajectories, each of which represents the development of some aspect of the originating condition (CE 141–2). These aspects existed first in a state of interpenetration in a primitive tendency that preceded their dissociation across distinct lines (CE 116). Divergent evolutionary trajectories may appear complementary in certain respects, owing to the fact that the qualities that define them were previously unified (CE 117, 135). I return to this idea nearer the end of the book.

In Chapter 1, I explained Bergson's rejection of the idea that everything could be given in advance. Both actualism and possibilism remain committed to it. The modal status of tendency has to be otherwise. Tendency cannot be located at any determinate point in space and time. It is not actually anything, being only an inclination towards actuality. Tendency is pre-individual, pre-particularised. The forms that express it are neither possibly nor potentially specifiable in advance. The directions it will take over the course of some particular trajectory can be thought to inhere within a tendency only after they are dissociated from it, that is, only retrospectively. Tendency is 'the forward thrust of an indistinct multiplicity', the modal register of which – as I discuss below – is virtual (TS 293). Tendencies are actualised progressively, along ramifying lines. They are individuated in their interaction with inhibiting factors, such as material resistance (CE 258). Tendencies respond to those inhibitions by splitting up and branching across a set of whatever directions are available and advantageous. Each direction is then defined by the tendency that has predominated within it by suppressing the tendency that opposes it. The tendency suppressed in one direction is predominate in the other. In this way each tendency is allowed to develop along a trajectory specific to it (TS 297). Yet the impetus, effort, or energy sustained along each line of development – its principle of activity, the reason for its tending onward – is derived from the original tendency, the original multiplicity from out of which it was dissociated (CE 119). Each individual tendency contains within itself not only the tendency that opposes it in a state of passivity, but the original multiplicity in a state of virtuality as well.

Matthias Vollet understands Bergson's mature conception of tendency on the basis of ceaseless movement, auto-differentiation in two directions, interpenetration of elements, developmental open-endedness, and arborescent particularisation (2008: 291). Elsewhere he emphasises the characteristics of directional unpredictability, and complementary or antagonistic composition (2012: 363). I think that is mostly right, though it misses some of the mereological features of tendency, such as the contraction of the whole in its parts. On my view, the idea of tendency can be summarised as follows. Tendency is directional movement, inclined towards actuality. The tendency towards actuality is multiple, dense with heterogeneous inclinations tending in a number of directions. Multiple tendencies are like multiple psychological states, blending into and informing each other. Particular tendencies branch away from that original multiplicity, but retain their continuity with the rest in the form of the history that they carry with them as they develop. Directional orientations run across a spectrum from vitality to materiality, freedom to physical determinism, creativity to repetition. Particular tendencies tend towards some limit on that spectrum, which provide them with immanent finality. What will become of a tendency cannot be predicted because its development is open-ended and contingent. Its ability to attain its limit is inhibited either by competing tendencies towards actuality or by the medium in which it materialises. Tendencies are not necessarily terminated or dissolved by inhibition, tending instead to divide themselves around obstacles, exploring new trajectories in order to extend themselves further, like a river winding around the rocks that block its path.

The modal-mereological difference

Tendency is the actualisation principle of virtual existence. Virtual wholes tend towards their actualisation in parts. Tendencies are unified multiplicities, virtual wholes themselves, distinct and continuous, each particular tendency interpenetrating with others in a series. These are the features of qualitative multiplicity. The modal status of such multiplicities is virtual. Earlier I explained how virtual instantiation allows for the inherence of a whole in each of its parts. Quantitative multiplicity, or the externalisation of interpenetrating elements, functions as a filter for the division and determination of a virtual whole. It is not so much an independent and unrelated register as it is an operation performed upon – a conversion, a transformation of – qualitative multiplicity, whether conceived psychologically as the fundamental datum of inner experience or ontologically as tendency.

The virtual is obviously different from the possible, virtual wholes having little in common with possible worlds or sets of possible things. Virtual tendencies are also different from potentialities. According to the Aristotelian conception, potentiality is defined either by an identity with its actualisation (as with possibility), or else by its telos, the end to which it is oriented and by which it is individuated. 'We speak of perceiving in two ways', writes Aristotle, 'for we say that something sees or hears both in the case of something that has the potentiality for seeing or hearing, even though it is asleep at the time, and in the case of something that is actually seeing or hearing at the time' (1995: 2.5; 417a10–13). This is the identitarian distinction between potential and actual: the closed eye sees potentially, the open one actually. Potentiality refers, in the first sense, to the horizon of identity a dormant capacity shares with its actual manifestation. While the difference between possible and actual is one of spatial location, understood in this way, potential differs from actual only in terms of employment, application, or exercise.

Aristotle affords it a differential definition as well. Taking the way we speak of knowledge as a model, he distinguishes two ways by which potentiality is actualised. We say, in one sense, that 'a man knows because man is a kind of thing that knows and has knowledge;' and, in another, that a man 'who has grammatical knowledge knows' (1995: 417a23–6). To know the rules of grammar is to harbour an identitarian potential for knowledge. It is already to have the knowledge that demonstration actualises. 'But in the first case, we [pass from potential to actual knowledge] by being altered through learning, and by frequent changes from the contrary state' (1995: 417a31–2). This is potentiality freed from the strictures of resemblance, as 'alteration' and 'frequent change' denote a difference between the potential for knowledge and its actual products. Yet this difference comes with teleology. We are altered in the process of learning, but this is end-directed change. It is guided, and its aim is present already before learning occurs. If the subject has the potential to know before learning, this is because the acquisition of knowledge 'is a change into possession of a state and into the fulfillment of the subject's nature' (1995: 417b16–17). This state is not prefigured before its actualisation (as in identitarian potentiality or possibility), but the potential for its possession is nevertheless oriented towards it in advance. The potential for a state is the specific capacity to attain that state, a capacity, or tendency, structured by a determinate disposition towards it. The answer to what a potentiality is going to become can be given prior to the becoming. The potential contributes nothing new, nothing creative or dynamic to that process.

Not so for virtuality. The virtual neither resembles nor prefigures its actualisations. It is pre-individual, pre-particular. If it is oriented towards anything more than particularisation or actuality, it is only vaguely so, by being inclined in the direction of novelty or homogeneity. Directional determination takes place progressively, as tendencies develop and get closer to the actuality of some actual form. Developing tendencies are something like potentials, but they are dynamic, creative, open to a number of developmental outcomes, continuous with other tendencies in a series, and carrying the history of their dissociation from that series with them as they progress. Actualisation involves the dissociation of tendencies from out of their mutual implication and their individuation along lines specific to each of them before it results in the concrete spatiality of material forms.

The difference between the actualisation of a possibility and the actualisation of a virtual tendency is in the fact that the former process consists only in a substitution in modal register while the latter contracts the virtual whole from out of which it was dissociated by instantiating that virtual whole in a spatialised part. In his monograph on Bergson, Deleuze writes that 'what coexisted in the virtual ceases to coexist in the actual and is distributed in lines or parts that cannot be summed up, each one retaining the whole, except from a certain perspective' (2006: 101). The difference between virtual and actual is neither addition nor subtraction (as with possibility), nor is it realisation and completion (as with potentiality). It is best conceived in terms of contraction instead.

Tendency should be understood as split across the indistinct directionality of a virtual-whole and the series of determinate directions taken by the individual tendencies dissociated from it. The virtual-whole of tendency requires that (1) each of its directions interpenetrate, (2) the state of their interpenetration is organised in terms of a whole, and (3) any one tendency instantiates the whole in which it is implicated. Particular tendencies are tendencies for which (1) each direction is realised by subordinating the direction that conflicts with and inhibits it, and (2) each is external to the other, extending itself as far as it can along its own line. The third component is retained by particular tendencies. This mereological retention of the virtual-whole from out of which individual tendencies are dissociated is indispensable to Bergson's philosophy of evolution. The virtual is efficacious in and through its particularised parts because it is mereologically instantiated in them.

The actual, spatially distributed, is always a bringing-into-actuality of what exceeds it. There is always more to be drawn into actuality, even if every particular actualisation instantiates the whole by determining it through a set of parts, psychological states, images, or actions. Time overflows its spatial determinations. Potentials cease to be potential once they realise their ends. A seedling is only potentially an oak tree before actually becoming one. Tendencies are not like that. There is always more that they can do, more and different development to be explored, other virtual elements to be drawn into actuality. Tendencies are never fully or finally exhausted because the future does not exist, the whole is open, there is always a new direction that they can take. Tendency development is like evolutionary variation. It cannot be forecast ahead of time and it does not stop happening.

Virtuality and the dispositional modality

Bergson's theory of tendency had a number of important predecessors in the history of philosophy (see Kistler and Gnassounou 2007: 2–27; cf. Anjum and Mumford 2018: 24–46). Many of its principal commitments continue to attract advocates in the metaphysics of causation as well. Without referring to Bergson, Rani Anjum and Stephen Mumford have recently undertaken to rehabilitate a notion of tendency in the metaphysics of powers. They propose that all of causation is fundamentally tendential (or dispositional), and that tendencies are neither possible, nor necessary, nor contingent, but modally *sui generis*. Like Bergson and the non-mechanistic tradition in French philosophy of nature that he continues, Anjum and Mumford contend that this irreducible 'dispositional modality' is intrinsic to interior life, given in the experience of agency, and that it is on that basis, by analogy, that a metaphysical account of tendencies can be elaborated (see Sinclair 2015, 2019: 183–211). I conclude this chapter by introducing their argument, parsing Bergson's theory of tendency in their terms, and then demonstrating what it is that sets Bergson's view apart.

Anjum and Mumford endorse a position they call causal dispositionalism (2018: 9). The idea is that all causation is dispositional, a matter of dispositions manifesting effects. Disposition means the same thing as power, potency, potentiality, or capacity: the ability to produce an outcome without necessitating it (2018: 8). The relation between a disposition and its manifestation is not one of necessitation, but neither is it one of contingency or mere possibility. It is a matter of tendency, a causal power halfway between contingency and necessity. 'A cause tends or disposes towards its effect' (2018: 9). This means, first, that a cause 'can sometimes succeed in producing it', the effect, if the conditions are right, though nothing necessitates the production. Second, there is an inclination towards that particular effect, distinguishing it from other possibilities and making the effect likely (2018: 9). A billiard ball, when struck, will tend to move. Nothing necessitates that it does, as not all struck balls do. It is also possible for the ball to vanish, as Hume suggested, but vanishing is not a tendency of billiard balls. There is a difference between movement as a mere possibility, which is shared by basically everything, and the tendency towards movement, which is a power of particular things, like billiard balls. All causal relations are relations of this sort, relations between tendencies and what they tend to do. This idea makes for a 'deeply tendential' view of nature (2018: 11).

It should be easy enough to accept that causation involves something more than mere contingency. We should be able to explain why caffeine tends to improve focus. But why deny that causes necessitate their effects? Why is tendency not just a matter of necessitation under the right conditions? There are two reasons, corresponding to Anjum and Mumford's two principles of tendency, internal and external. The external principle holds that causes can always be interfered with or counteracted in a way that prevents their effects from occurring (2018: 11). If causation were necessitation, then it should be impossible to have a cause without the typical effect. Every case of the cause should be a case of its effect. Interference examples prove that wrong.

Philosopher of science Nancy Cartwright is perhaps the most popular advocate for the external view, though she continues a line of thinking about causation that dates back at least as far as Mill (see 1882: Book IV). Cartwright prefers the term 'disposition', which she uses to refer to capacities whose manifestations occur unless they are interfered with by the manifestation of other capacities (1999: 73 and 2007: 203–5). That capacities can be inhibited is what marks them out as capacities. That they are disposed towards manifestation in the absence of conflicting capacities is what makes them dispositional.

Interference comes in two forms. Subtractive interference takes something away from a cause, as when one removes the oxygen keeping a candle burning by covering it with a jar. Since subtractive interference prevents an effect by removing part of its cause, it does not suffice to undermine necessitation. The argument against necessitation requires that a cause is present, fully present, but without its typical effect. This is supposed to be the case with additive interference. Additive interference adds something to a cause that prevents the manifestation of the typical effect. Instead of removing oxygen from a burning candle, one could wet its wick. Arsenic can be neutralised by an antidote. Forces can be counteracted by opposing forces. The idea is that 'a cause, which is capable of producing a particular kind of effect, can be prevented from doing so because of some factor that is external to the cause; and it is this external factor that an additive interferer is adding' (Anjum and Mumford 2018: 12). That is the external principle of tendency. It shows that 'causes did not necessitate their effects even on the occasions where they succeeded in producing them' (2018: 11).

There are two problems with the external principle. The first is that it appears reducible to necessitation under the right conditions, amounting to the 'conditional necessity' view of causation (2018: 13). This view holds that a causal power will necessarily produce its typical effect unless a counteracting power obstructs it. Causation is necessitation, conditional upon the absence of interfering factors, with no other modal category required. Anjum and Mumford deny that the external principle can be reduced to necessitation in this form. Conditional necessity assumes necessitation as the default modal category. It explains away undermining factors as wrong conditions to save the assumption that a cause will lead necessarily to its effect in principle. One might propose that the ingestion of an antidote after arsenic does not break the causal necessity of arsenic so much as it makes for a different causal situation, a different causal whole, one that necessitates a typical effect of its own. But that is a just-so story, one that saves the necessary relation between a cause and its effect by explaining every change in the effect as the still-necessary result of a different composition of causes. Conditional necessity appears to be insulated from falsifying scenarios. There are other motivations for the view, and other counter-arguments against them (Anjum and Mumford 2018: 13–17). Anjum and Mumford conclude that since 'there seems always to be the possibility of additive interference that can stop an effect from happening', it is reasonable to maintain 'that causes no more than tend towards their effects' (2018: 17).

The second issue with the external principle is that, supposing it cannot be reduced to conditional necessity, it appears to offer little more than that. It says that causes do not necessitate their effects. Causal power involves another modal category, a dispositional modality. But the external principle does not tell us what it is, or anything about it, and so appears to lack positive content. Mark Sinclair was the first to make this objection (2015: 166–7, referring to Mumford and Anjum 2011). Anjum and Mumford respond by claiming that the external principle of tendency also contains a directedness component, which specifies what it is that a tendency tends to do, absent prevention (2018: 13). Sinclair insists that that idea of directedness merely restates the idea of force as conditional necessity, and so fails to add any positive content to the idea of tendency or its dispositional modality (2019: 206).

On Anjum and Mumford's account, 'the external principle contains the positive component of a cause that typically has an effect', sometimes called 'the directedness component of the dispositional modality' (2018: 13). Directedness means that causes are disposed towards their effects; they are *for* them. But this does not address the concern, which targets the modal status of that direction. If causes are directed towards their effects only in the sense that they do not necessitate them, then in what sense do they bring them about? In what, other than non-necessity, does the dispositional modality consist? 'Less than necessity and more than pure contingency' cannot be the whole story (2018: 13).

Anjum and Mumford succeed in defusing the objection that the interference model for causal power collapses into conditional necessity by introducing a second, internal principle of tendency. The internal principle does not refer to external factors in its conception of causation. Causation is not necessitation, but not only for counter-factual reasons, because a cause could always be counteracted. A cause need not bring about its typical effect even when all the conditions are right. According to the internal principle, 'there is nothing that prevents the effect from occurring, but still, it need not occur, just because the modal nature of the cause is internally tendential' (2018: 18). This principle makes tendency irreducible to conditional necessity. A cause tends towards its effect in that it typically produces that effect (or set of effects) in particular and not any effect or set of them whatsoever, *and that it need not produce that effect*, even when there is nothing that prevents it, even when ostensibly suitable conditions obtain. This view commits Anjum and Mumford to a rejection of the Principle of Sufficient Reason, as effects are intrinsically underdetermined by their causes. Underdetermination does not mean that anything goes, as on some variant of a Humean model, for different tendencies dispose towards their effects with greater or lesser strength. Strong tendencies might be seen to manifest reliably, while weak tendencies may only manifest their effects some of the time. 'Rather than a matter of chance, then, the occurrence of a tendency's manifestation is produced by that tendency or causal power' (2018: 18). The production of an effect is irreducibly tendential, which means that the relationship between a tendency to produce an effect and the production of the effect is itself a matter of tendency as well.

There are three notions of tendency at play here. According to the first, the producer of an effect, the causal power or potential for that effect, is a tendency. A magnet has the tendency to stick to a metal in that it has the power to, and may manifest that power depending on the circumstances. There might be no such manifestation, but the tendency is there in the magnet nevertheless. A second sense of tendency refers to the effects that tendencies produce when those effects take the form of statistical regularities. Burning is a tendency of lit matches. It is an effect of a match's causal power or tendency to burn. The repeated occurrence of burning can be called a tendency as well, a statistical tendency. The third and final sense of tendency refers to the modality of the production itself. When a struck match burns, it realises its tendency to do so 'in a way that is less than necessity but more than pure contingency' (2018: 22). Powers are tendencies in the first sense, and their modal status – the relationship that obtains between a producer and its product – is tendency in the third. The third sense of tendency is irreducible to conditional necessity. It is what Anjum and Mumford call the dispositional modality.

The dispositional modality consists in a non-necessitating directed-ness towards some set of causal outcomes, specifiable to some degree of strength or reliability (see Glynn 2012: 1102). This modality is supposed to characterise all powers. Anjum and Mumford think that it is easiest to locate in ourselves, in action. The principle of epistemic access to the dispositional modality is experiential. It is 'known directly from experience' (Mumford and Anjum 2011: 195). This facet of Anjum and Mumford project involves a counter-argument against Hume's scepticism regarding the experience of causal power in agency (see Sinclair 2019: 190–2). According to their 'reunificationist' response to Humean scepticism, we have a direct experience of our own agential powers, our ability to realise intentions in actions, to will that they happen, and to bring about a predictable set of effects in the world as a result (Mumford and Anjum 2011: 212). Agential power always has a 'limited class of outcomes, out of all those that are merely possible' (2011: 210). The manifestation of agency in some such outcome is never a matter of necessity, as any power can be inhibited or prevented from manifesting itself. This, too, we know directly from experience (2011: 212). Agency is tendential, evidencing both 'directedness (with a certain intensity) and the possibility of prevention' (2011: 210). It is on the basis of this inner grasp of tendency or disposition that an account of the dispositional modality of causation can be constructed by extension.

Bergson's theory of tendency involves elements of Anjum and Mumford's account, but differs on the nature and actualising operation of the modality at stake. On Bergson's conception, tendency is irreducibly durational in each of its facets. That is what sets his view apart.

Bergson holds some form of each of Anjum and Mumford's two principles of tendency. The external principle states that causal powers

do not necessitate their manifestations because they can always be counteracted by the presence of conflicting powers. Causation is tendential
because it is susceptible to interference. Bergson subscribes to something
comparable. His view is that tendencies tend towards actuality, tending to realise themselves as far as possible. Interference from conflicting
tendencies or constraints in the medium for their realisation is what
keeps them tending. If there were no conflict, no constraint, then the
tendency would be all that it is already and would no longer have to tend
towards anything.

Can Bergson's position be reduced to conditional necessity? Do
tendencies necessitate their effects conditional upon the absence of
interfering factors? The idea is misguidedly abstract. There is no such
thing as a single tendency, tending towards actuality on its own, in isolation from others. To imagine it that way would be to imagine a fully
realised property, state, or thing, full stop. What defines a tendency is its
interaction with others. Interaction in the domain of tendencies is different from interaction between disparate individuals. Tendencies are not
independently definable, each in competition for an individual end that
could be specified in advance. The conflict and composition are primary.
Tendencies interpenetrate, fading into each other on a continuum, in a
whole, like psychological inclinations. Acting on one requires suppressing others, and even then, the actualising inclination borrows from the
others, is suffused by them, and can be considered one individual thing
only after the fact, from the perspective of its manifestation.

Bergson's variation on the external principle locates interference internal to the nature of tendency. His position on Anjum and Mumford's
internal principle would have to depend on how the principle is formulated. If the idea is that tendencies do not have to manifest their effects
even in the absence of conflicting factors, then again it is a mistake. There
is no tendency that tends in the absence of conflicting factors. Limitation
is hardcoded. Perhaps the internal principle could be taken to state just
that tendencies always underdetermine their effects. The latter cannot be
assumed to follow necessarily on the basis of the former and in advance.
If that is the idea, then Bergson would affirm it. Underdetermination is
a temporal fact, a consequence of the non-existence of the future. The
internal principle might be thought in terms of the distinction between
prediction and explanation, which I mentioned in Chapter 1. The outcome of any tendency cannot be predicted before it occurs. In this respect,
tendencies underdetermine their effects and the relation between them
is not one of necessity. Bergson agrees with Anjum and Mumford that
underdetermination does not imply complete contingency, as if a given
tendency might lead to any outcome whatsoever. A set of effects can

always be explained on the basis of the causes that produced them. There is a rational coherence between the two. The stipulation is that explanation is necessarily retrospective. The Principle of Sufficient Reason operates on effects that have already taken place, not on tendencies prior to their manifestation as effects.

From a Bergsonian perspective, Anjum and Mumford's position on the relationship between tendencies and their effects is ultimately unsatisfactory. They do not distinguish between tendency, power, disposition, capacity, or potential. A tendency is a tendency to manifest some particular effect. What makes a tendency a tendency and not a necessitating cause is the absence of necessitation. But tendencies can be specified in terms of the effects towards which they tend nevertheless. In this respect, they are just like potentials, whose orientation is defined in advance by the actual ends whose potentials they are. Anjum and Mumford add inclinational intensity to the idea of potential, so that a tendency is a potential to produce some effect to some degree of intensity, being more or less likely to manifest itself. One wonders, then, if the end can be given in advance, what it is that prevents a potential from being actualised in the absence of conflicting factors. If tendency is potential, already oriented towards the effect that it will produce, then what is it that accounts for the underdetermination?

Anjum and Mumford admit that they do not have much to say by way of argument for the internal principle (2018: 20). They do not provide a metaphysical account for it. Instead they seek to demonstrate the explanatory utility of the principle by showing what problems it resolves if it were to be accepted (2018: 23). Bergson's account of tendency as non-necessitating directional inclination is, on the contrary, a metaphysical one. As I explained above, it involves a distinction from potentiality. Tendency in Anjum and Mumford's first sense of the term is power or potency taken independent of its exercise. Bergson affirms the reality of unrealised tendency, but denies that it is equivalent to the potential to bring about some particular effect. Tendencies are continuous with other tendencies before being realised. Primary conflict and original competition are what distinguish tendencies from potentials. Tendencies can be isolated, identified, and correlated with their effects only after those effects occur. The actualisation of a tendency in process is always at least minimally creative. If tendencies underdetermine their effects it is not only because they do not necessitate the occurrence of those effects, but because they cannot be defined in terms of their effects in advance. Directional inclination is non-necessitating in just that sense.

Anjum and Mumford distinguish the first sense of tendency as potential from tendency as a modal register of its own. Bergson affirms that there is an irreducible modality of tendency as well. He calls it virtual. Like the dispositional modality, virtuality exists between necessity and contingency. It is dispositional in a sense that sets it apart from inert possibility. Yet it retains an unpredictability, which is Bergson's concept for what Anjum and Mumford try to capture with the idea of underdetermination. There are two salient differences between Bergson's virtual and the dispositional modality. The first is the one I mentioned above, the interpenetration of tendencies, their original conflict. For a tendency to be modally dispositional means that it need not manifest an effect even though it tends to. For a tendency to be modally virtual is for it to be part of an interpenetrating whole, identifiable with a particular effect only after the effect occurs. The second difference is a temporal one. It concerns manifestation as a process. The virtual is not a static repository of distinct tendencies. It is in time and should be conceived according to the features of duration. It retains its history of development, contracts its past in present form, and tends towards a variable future. The manifestation of a tendency is part of the history of that tendency, altering how it can be made manifest in the future. As tendencies interpenetrate, they alter each other as each one develops. This is also why tendencies cannot predict for their effects. It is not only because they are modally tendential, as Anjum and Mumford contend. More informatively, it is because the modality of tendency, the virtual, is temporally structured. Novelty is a defining characteristic.

When Anjum and Mumford model the dispositional modality on the psychology of action, they continue a tradition in nineteenth-century French thought of appealing to the subjective experience of agency in thinking about force. It is also Bergson's tactic (see Sinclair 2019: 163–82). I have shown how Bergson elaborated his own theory of tendency on the basis of his conception of inner life and the expression of free acts in *Time and Free Will*. The difference is in Bergson's account of what inner life is like. Each of the other differences between the two views is a consequence of this one. For Bergson, we know ourselves more profoundly than we can hope to know any other object. The content of that self-knowledge is ceaseless change (CE 1). We experience ourselves via change of state between sensations, feelings, volitions, and ideas, and we experience each of those states as ceaselessly changing in themselves as well. This is because everything we experience is coloured by memory, which 'swells with the duration which it accumulates' (CE 2). Each of our states 'continue each other in an endless flow' and so cannot be considered distinct (CE 3). Cognition and perception succeed in introducing discontinuity into this flow for practical purposes. But in desiring, willing, and acting, 'our past, . . . as a whole, is made manifest to us in its impulse . . . felt in the form of tendency' (CE 5). Each ostensibly distinct act contracts the whole history of our character. Action is then added to that history, modifying it in turn. The result is that 'we are creating ourselves continually' (CE 7). The future is kept open and unpredictable as the past continues to develop in being made manifest differently. Bergson might agree with Anjum and Mumford that tendency is a psychological datum in agential experience, but he would no doubt object to their account of agency for its erasure of the reality of duration. Properly conceived, introspection *can* furnish the basis for a modally irreducible theory of tendency, one that can be extrapolated in order to generate a 'deeply tendential view of nature', but interpenetration, unpredictability, contraction, and novelty should be considered ineliminable features of it.

Conclusion

At the core of Bergson's metaphysics is the idea that time is real. Its reality, which consists in the ongoing retention of a past opened onto a dynamic future, eludes the modal categories of the actual (or necessary), the possible (or contingent), and the potential (or dispositional) alike. Another modal category is required. The virtual is that category. Tendency is its actualisation principle. Seen from this perspective, tendency is one of the central concepts of Bergson's philosophy of time. It is modelled on the analysis of effort, and worked out through the whole range of his texts. Reading across them allows us to discern the outlines of a theory of tendency in development. What originates as a psychological fact disclosed to introspective analysis, setting lived duration apart from spatial phenomena, culminates in a multi-faceted view regarding the nature of reality and the essence of evolution, uniting inner life with cosmological becoming. It is not just our characters that are felt in the form of a tendency to act, contracting a past history of development in open-ended impetus. Reality itself is tendency at the most fundamental level.

The idea of directional inclination, tending towards a limit, is the key to a number of important concepts and distinctions in Bergson's philosophy of evolution. In the next chapter, I show how tendency is at work beneath the life/matter dyad. The rest of the book develops an interpretation of *Creative Evolution* on that basis.

4

Individuality and Organisation

This chapter reconsiders Bergson's views on space and time on the basis of his metaphysics of tendency. Space and time are reified abstractions from spatialisation and temporalisation processes. Life and matter are not categories under which any actual entities could fall, but principles that define a set of conflicting tendencies towards time and space. All that exist in fact are composites formed out of the configuration of these tendencies under differing relationships of predominance and subordination. Organised material systems are typified by the tendency towards time. The *élan vital* is an image for the individuating activity of the temporalising tendency.

Introduction

Bergson seems to accept a basic duality between life and matter when he defines life as 'a tendency to act on inert matter' (CE 96). Life is action that 'always presents, to some extent, the character of contingency; it implies at least a rudiment of choice' (CE 96). Matter is the medium for the activity of life. Without life's 'effort to engraft on to the necessity of physical forces the largest possible amount of *indetermination*', matter is inert, unorganised, its behaviour determined by laws (CE 114). When Bergson makes claims like these, he is speaking abstractly. There is no such thing as fully inert matter on the one hand, absolutely extended in a homogeneous space, set against completely liberated forms of life on the other, wholly free from the causal laws governing physicochemical forces. Life and matter are principles abstracted from tendencies. Tendencies can be oriented towards time or space. Time and space refer to opposite directions on a continuum. To tend towards time is for something to tend to emphasise the efficacy of time, which is manifest in processes such as self-organisation and the interpenetration of parts in a developmental whole. To tend towards space is for something to tend to approximate the abstract ideality of a geometrical grid on which its parts can be isolated, localised, and modelled deterministically. Life itself exists on its own no more than inert matter. To refer to life that way involves two operations of abstraction. It involves an abstraction of fully realised states – of self-organising individuality, interpenetration, and indetermination – from the tendency towards their processual development. It involves an abstraction of this tendency from the material systems in which it is embodied at some degree of realisation, in conflict with the tendency towards space as well.

There are two registers on which these ideas play themselves out: the individual, where life is defined by the tendencies active in the organisation of organisms; and the evolutionary, where life is defined by the directionality of the processes of variation and speciation. Life, in both registers, is Bergson's name for a state of affairs in which the tendency towards time is predominate over the tendency towards space. That predomination is what explains both organisation and evolution alike.

In this chapter, I treat Bergson's conception of time as an individuating tendency. I turn to the evolutionary register in the next. In order to make sense of time, I begin with space. The first sections of the chapter parse the spatialising tendency according to three facets: (1) the isolation of a material system from its interactions, (2) the externalisation of its parts from their mutual co-implication in a whole, and (3) the localisation of the system's parts across an abstract grid. I develop the idea of time as an individuating tendency by inverting the three facets of spatiality. Organisms are defined by (1) the thermodynamic conditions and properties of organisation, (2) the mutual implication of elements in the developmental whole of the organism, and (3) the correlation of level of duration with the metabolic activity of self-organisation.

Spatialisation

a. Isolation

The first definition of matter in *Creative Evolution* involves a tendency towards isolability (CE 10). In Chapter 2 I mentioned that unorganised matter forms a complex whole. Matter depends for its subdivision into determinate bodies on an individuating principle. That principle is perception, which fashions actionable images by individuating matter according to the practical dictates of action. The material systems studied by the physical sciences need to be individuated as well. These are systems whose composition, structure, properties, and behaviour are supposed to be objective, which means that their determination cannot rely on pragmatic perception alone. Material systems are also constituted by elements that elude the thresholds regulating the perceptual apparatuses of those who study them. Material systems have to depend on another individuating principle. That principle is isolation. If perception divides matter into bodies, then science isolates matter into systems (CE 12).

Mechanistic science individuates its objects of study by isolating them from relations to other bodies and processes, delimiting their boundaries and closing their causal activity from interruption or interference from external influences. This operation of isolation and closure constitutes a material system by setting it apart from everything outside it. Only with respect to a boundary distinguishing what is internal from what is external and marking the closure of the former from the latter can a system can be mereologically decomposed into its constituent parts, each of which is then isolable and decomposable in turn. As these parts are ultimately understood as changeless and indivisible units whose rearrangements account for the changes undergone by the system as a whole, the system's future states can be modelled and predicted in advance, at least in principle and by a sufficiently powerful intellect (CE 8). Bergson holds that this kind of predictability requires the erasure of the reality of time. The fully isolated system is also mechanistically tractable and deterministically describable. Isolation serves then, as a pre-condition for the articulation of laws of nature as well (see Vieillard-Baron 2008: 576–80).

There may be something unsatisfying about this conception. Systems science recognises three categories of system, only one of which it designates as isolated and any of which may or may not be deterministic (see Bertalanffy 1968, Nola and Irzik 2005: 376, and Gitelson et al. 2003). Bergson seems to equivocate between the individuation and isolation of a material system, mereological closure, and the laws that describe its activity. I suggest that isolability is best understood as an abstract end that defines and orients a tendency that is realised in each kind of system to a different degree.

Closed and isolated systems are both defined against the matter-energy exchanges that constitute open systems. An open system is a system that exchanges both energy and matter with its environment. Most ecosystems are open in that they are (1) patterned by interactions between the organisms that compose them and external influences; as well as (2) regulated by changes in an external energy source, like the sun. Closed systems are systems that are materially self-contained, but still open to their environments energetically. Ecologically closed systems are difficult to locate; most are artifacts either of idealisation or modelling, or else of human manufacture. A self-maintaining aquarium is one example of a human-made closed system, isolated from the introduction or excretion of material elements. Yet it remains open to energy flow from its environment in the form of external lighting, heating, the electricity required to run its filtration system, and so on.

Isolated systems are defined by their closure against the exchange of matter and energy with their environments. Bergson holds that these kinds of systems are ultimately impossible to locate in the material world without either abstracting away from the minor external influences that open all systems to their outsides (by modelling them), or else by fixing the limits and conditions of the system and so instituting its closure artificially (by experimental control) (CE 10). Most isolated systems are isolated through a combination of these strategies. For instance, the rates of reaction between chemical processes in the closed vessels of physical chemistry are expressible in terms of elementary laws and describable in terms of mathematical models because interactions with external processes are fixed (see Espenson 2002: 254–6). Isolation is therefore always only ideal.

Bergson equates complete isolation with determinism (CE 8–10). In ideally isolated systems, causal processes are governed by necessary laws (CE 210–17). As those laws describe and predict everything that has and will occur in the system, a sufficiently powerful intellect could map each of the system's changes in advance (CE 8). Time has no significance, rendering the system deterministic. Yet the equation of isolation with determinism does not hold for general systems theory. According to the theory, deterministic systems are inclusive of the open, the closed, and the isolated without being necessarily entailed by any of them. Whether a system is open, closed, or isolated does not tell us whether or not it is deterministic. A system is deterministic if and only if its state at $t2$ is necessarily entailed by the state it was in at $t1$, and so too for all of its states at all points in time. The outputs of a deterministic system are necessary consequences of the inputs. Open systems can be deterministic if their outputs are entailed by their inputs independently of changes in the material and energetic interactions to which they are open. If the relation between input and output varies with fluctuations in its matter and energy exchanges, then the system is either probabilistic or indeterminate. But its openness does not specify for its inclusion in any of these causal categories. The same point holds for closed and isolated systems for the same reason.

Bergson's equation of isolation with determinism is a result of his belief that science is founded on a set of mechanistic intellectual habits that function by suspending the significance of time for their objects of study (CE 195; 329–43; cf. Vieillard-Baron 2008: 583–6). If the scientific individuation of matter operates according to matter's isolability, and if that isolability – taken at its ideal extreme – requires the denial of the reality of time, then unorganised material systems are deterministic systems as well.

We should ask first why and on what grounds science should be said to individuate matter through the constitution of isolated systems; and, second, why and on what grounds isolated systems require the suspension of the significance of time for the description of their behaviour. Answering these questions allows us to coordinate Bergson's ideas about science and determinism with his ideas about matter, time, and space.

When Bergson defines matter by its tendency towards isolability, he means that the operations by which science constitutes, closes, and isolates its systems of study are not artificial, arbitrary, or externally imposed. Isolability is indicated in matter itself. It is not fully realised there, for otherwise it would not be tendency but property. Bergson holds that matter forms a continuous flux in which every point exerts some influence on every other and is shaped by every other to some degree of relevance (CE 203, 188; MM 292). As that continuum is not homogeneous but energetically differentiated, we should expect some of its subregions to be more or less isolable from their relations to the regions around them (MM 266). Scientific mechanism seizes and elaborates upon this isolability. Since their isolability is suggested in matter, material systems can be more or less appropriately constituted, and more or less realistically closed.

If 'matter has a tendency to constitute *isolable* systems', it is nonetheless 'only a tendency. Matter does not go to the end, and the isolation is never complete' (CE 10). Complete isolation is accomplished through the elaboration of the concrete tendency towards isolability to its terminal realisation in an ideal state. That state can only exist in abstraction from the material continuum discretised by it. If the isolated system is the *principium individuationis* of unorganised matter, then it is only inasmuch as isolation designates matter's end. Principles define and orient tendencies, but can be realised as states only in abstraction from the concrete reality that those tendencies qualify. To say that matter is individuated in the isolated system is to say that matter embodies a tendency towards individuation that is only able to realise itself as a complete property or state in the isolated system. As concrete material systems are only ever more or less individuated to the extent that they are more or less open or closable, and as complete isolation designates the terminal pole of the closable system, isolability operates as the end towards which the other kinds of systems tend.

Far from collapsing open, closed, and isolated systems into each other – and equivocating between each of them and the deterministic system in turn – Bergson should be understood as reconfiguring each category of system as a phase in the realisation of the same tendency. The open system, structured by the suggestion of isolability but in both material and energetic exchange with its environment, is the first phase in this tendency's progress towards completion. The closed system, materially self-contained but energetically open to its environment, is the second phase in this progress and probably the limit case of how far concrete matter can individuate itself. The fully isolated system is its end.

The second question I asked above is why and on what grounds the isolation of a system should be thought to require the suspension of the significance of time in its description. The isolation of a system is what permits mereological decomposition. It is at the level of its parts, once they are determined independently of each other and of their implication in the whole, that the isolation of a system corresponds with its availability to deterministic description. Bergson defines this dimension of inert matter in terms of the 'perfect externality of parts in their relation to one another' (CE 203). He draws the concept of externality from the two multiplicities of *Time and Free Will*. While qualitative multiplicities include their parts within themselves in a state of mutual interpenetration, quantitative multiplicities dissociate those parts from out of their implication in the whole, externalising them. Externality is the second of the three facets of spatialisation (see Costelloe 1913: 144 and Barr 1913: 648).

b. Externalisation

Isolability conditions the mereological decomposability of matter because material systems can be decomposed into their parts only on the basis of an individuating limit that defines what is internal to the system – and what can be specified as a part – against what is to count as its environment. Philosopher of science Rasmus Winther explains that it is isolation and specification, what he calls 'partitioning', that conditions mereological decomposability. He agrees with Bergson without knowing it: 'the partitioning frame is exactly what provides the interpretation of a perspective's kinds of parts, which, before specification, are left completely open, in a manner analogous to a logical variable' (2006: 476).

At the abstract limit of complete isolation, the system is determinable in the form of its parts and their activity within the system, closed both to matter and energy interactions with what is outside it. To understand its parts and their relations is tantamount to understanding the system as a whole, as there is nothing added by the whole to the description of its parts. Under ideal conditions of isolation, mereological decomposability implies the reduction of the whole down to its parts and its erasure as the real unity of the system. This is sometimes called 'mereological determinism' (see Rosenberg 2002: 116). The whole is a necessary causal product of the composition of its parts and is exhaustibly explainable through the study of them alone. The practical upshot of mereological determinism is that the system's parts are made available for independent theoretical determination. They can be relegated to different forms of experimental manipulation and distributed across distinct research programmes – physics, chemistry, geology, dynamic systems theory, and so on.

The externalisation of parts takes place along vertical and lateral axes. Lateral or horizontal externalisation is effectuated by the initial determination of the whole *partes extra partes*, while vertical or hierarchal externalisation consists in the subsequent decomposition of those parts into their constituents according to scale. The primary, lateral process of externalisation proceeds by dissociating a relational unity into the constituents of which it is made, separating parts from their relations to each other and determining each independently. The vertical operation proceeds downward, from the parts laterally distributed to the parts of which they are made until a fundamental level is reached at which no further decomposition is possible. The denial of time regulates this procedure by drawing it onward in the direction of the two facets that define the ultimate end of mereological reduction, indivisibility and immutability.

What Bergson calls 'analysis' – the method of decomposition and reduction – is defined by this tendency, to 'push the division . . . as far as necessary', and to 'stop only before the unchangeable' (CE 8). Mereological decomposition reduces the whole to its parts, and the changes that characterise the activity of the system at any one of its scales down to the displacement and rearrangement of parts on its smallest scale, parts that are themselves immutable (CE 7–8).

Immutable parts can be distinguished from each other only externally, by means of their location in a homogeneous spatial medium (see Rynasiewicz 1995). So when Bergson says that there is nothing to prevent the return of such a part to any of its previous positions within a system, he means that there is nothing *internal* to that part that would distinguish any one of its positions from any other (CE 8). The only changes that this part is capable of are changes relative to its location in a fixed space. As no difference is made to a part by this kind of change, it can be inferred that 'any state of the group [the aggregate of parts to which the systemic whole has been reduced] may be repeated as often as desired, and consequently that the group does not grow old. It has no history' (CE 8). Systems lacking histories – lacking time – are able to repeat and return to their past states because their states are distinguished from each other by means of the spatial determinations of their smallest components. It is possible in principle to represent everything of which this kind of system is capable – its state space – by mapping the possible spatial configurations of its parts on a plane.

For a system to be capable of this kind of mapping amounts to the conclusion that everything it 'will be is already present in what it is, provided "what it is" includes all the points of the universe with which it is related' (CE 8). This is a mereological explanation of what classical physics calls the reversibility of time (see Prigogine and Stengers 1985: 75). If there is nothing in the whole that is not in its parts – if the states of such a system are reducible to the spatial configuration of its parts – then 'the future forms of the system are theoretically visible in its present configuration' (CE 8). To be able to model in advance every change of which a system is capable is equivalent to subtracting time from it. The logic of the material system so determined is amenable to the simple law of mechanistic analysis: 'the present contains nothing more than the past, and what is found in the effect as already in the cause' (CE 14). Bergson concludes that 'all our belief in objects, all our operations on the systems that science isolates, rest in fact on the idea that time does not bite into them' (CE 8).

This is not so much a misprision as it is the elaboration of a tendency towards spatiality to its end, the reification of that end in the form of a state, and the treatment of that abstract state as if it were the concrete reality of matter. The tendency at stake is directed away from the significance of time and given in the form of repetition. If concrete matter never fully resolves itself in brute mechanical repetition, it is because matter – as I discuss in the next section – is at bottom composed of 'elementary vibrations, the shortest of which are of very slight duration, almost vanishing, but not nothing' (CE 201). The pasts and futures of these vibrations are immediate enough to be contained conceptually, or experimentally, within their present iterations. As those presents can be mapped onto spatial positions, time itself can be spatialised in turn, through its reduction to the measure of changes in the spatial configurations of a system's parts.

Bergson contends that the intervals between these spatial rearrangements do not figure into their calculation (CE 9). The flow of time that transpires across those intervals could assume an infinite or ideal rapidity, and 'the entire past, present, and future of material objects or of isolated systems might be spread out all at once in space, without there being anything to change' in the scientific formulae that measure them (CE 9).

These formulae work by attributing an 'abstract time t' to isolated material systems that 'consists only in a certain number of simultaneities or more generally of correspondences' between the states of the system and the points of a line along which they are distributed (CE 9). This number is supposed to remain invariant, 'whatever be the nature of the interval between the correspondences' (CE 9). Whether it takes Zeno's arrow five or fifteen seconds to reach its mark, so long as it were to pass (hypothetically) through the same points in space in each case, the scientific description of its trajectory should not change.

If this is an accurate portrayal of the classical physical description of time-reversible processes and if time is scientifically determinable in the form of a variable measure for the ideal quantification of a system's changes in state, then each change in state can be unfurled like a fan, all at once, coextensively (see Pilkington 1976: 15). This is spatialisation completed: the total externalisation of all parts, all states, all changes, from out of their implication in the systemic whole as well as from out of the duration across which they are progressively unfolded.

This is how we should understand Bergson's equivocation between isolated and deterministic systems. Isolation permits mereological reducibility, and as reducibility resolves itself in immutable minima, the macroscopic behaviour of the system as a whole can be described independently of the reality of time, as the rearrangement of parts in space. If a sufficiently powerful intelligence can predict the future states of such a system in accordance with the laws that govern the movement of its parts, then it is a deterministic system. If matter tends towards determinism as a concomitant of its tendency towards isolability, then it tends towards a mereological externality as well. This latter tendency conditions the scientific calculability of material systems, by making possible the treatment of each part independently. The complete externalisation of parts culminates, finally, in the localisation of each on a spatial grid. Localisation is the third component of spatialisation.

c. Localisation

Bergson adopted from Michael Faraday a precursor to what would become the quantum physical theory of matter (see Čapek 1971 and Beauregard 2016). On this theory, extended material bodies localised in a spatial medium are best represented as abstractions from smaller scale interactions of influence – fluctuations or excitations in a field – that are neither strictly localisable at any point in space nor given completely at any instant in time (MM 266, 292, 304, 313, 330; CE 188). The material world, independently of its perceptual articulation, is a decentred aggregate of points – relative centres of force or influence – each one of which 'gathers and transmits the influences of all the points of the material universe' (MM 31). Bergson comes to this position by way of Faraday's reconception of the nature of matter. Faraday imagines the atom as a point of compression out of which lines of force emanate in every direction, ultimately shaping every other material point in the solar system. Matter is best defined not in terms of determinate bodies composed of separable parts, but of '*modifications*, *perturbations*, changes of *tension* or of *energy*, and nothing else' (MM 266). The self-identical atom is an abstract postulate, derived and distilled from out of an energetically differentiated medium whose 'essential character . . . is continuity' (MM 259).

The Newtonian pronouncement that a dynamically energetic material continuum can be resolved into sets of self-identical atoms localised across a grid of determinate spatial points, fully present at determinate instants of time, is exemplary of what Whitehead called the 'fallacy of simple location'. Whitehead cites Bergson's critique of spatialised time as inspiration (1967: 64). The fallacy consists in the contention that a 'bit of matter' can be located absolutely 'in a definite finite region of space, and throughout a definite finite duration of time, apart from any essential reference of that bit of matter to other regions of space and to other durations of time' (1967: 58).

This is a fallacy, according to Whitehead's reading of the relativistic and quantum theories of matter, because 'all fundamental physical quantities are vector and not scalar' (1967: 177). Scalar quantities are quantities of magnitude, while vector quantities involve direction. Vectors are magnitudes directed to and from beyond the frame of any one of their localisations. The trajectory taken by Zeno's arrow is describable in terms first of vector; any magnitude attributed to it has to be constructed on the basis of its movement. Processes precede the phases or states out of which they can be considered composed. No unit of space or instant of time can be isolated from the overall 'physical electromagnetic field'. No material body can be completely extricated from its relation to the continuum of energetic interaction.

Neither can matter be abstracted from the temporal spreads that it is realised across. If space is a grid of simple locations, then larger scales can be straightforwardly divided into smaller ones – bodies into the protons and electrons that compose them, space into the smaller increments out of which the larger are made and at which each proton or electron ought to be located. But at the smallest scale, matter appears to be spatially discontinuous and temporally elongated. This is what Whitehead means by his contention that even electrons 'dance' (1964: 167). He derives this

characterisation from the quantum-theoretic inquiry into wave functions. Electrons, for quantum theory, are – like all quantum objects –
partially particle-like and partially wave-like, neither fully one nor the
other (Tonomura 1998: 7–13). Electrons are described by a probabilistic
quantum wave function with spatial spread, vibrating at some frequency.
Bound in a stable state in an atom, the electron wave function assumes a
shape called an 'orbital'. Orbitals do not contain electrons, but rather are
the electrons in this state. It is the orbital that vibrates, like an oscillating
three-dimensional wave. Electrons do not exist in any one locale, fully
manifest at any one instant. Electrons are processes, shaped by their relations within the bodies they compose. Those bodies – the macroscopic
entities that constitute the larger scale matter of classical physics – have
to be considered ultimately non-localisable as well.

Quantum non-locality refers to the fact that separate entities are, after
interacting and separating, still internally linked or mutually property-
constitutive in ways that do not register materially. At the quantum level
no location is independent of its internal relation with others, not just
because the bodies that occupy those locations are not fully manifest
at any one of them, but because the locations themselves are referen-
tially embedded in each other (see Epperson 2004: 85–6; 12–13). While
Whitehead endorses the view that every entity is constituted by its internal relations to the constituent entities of the entire cosmos, Bergson
holds that organised bodies are self-individuating in a way that affords
them at least some material independence.

Whitehead's metaphysical innovation consists in drawing from the
concept of process a spatial consequence in addition to the more familiar temporal insight. Processual phenomena manifest themselves over
a certain temporal spread. They are incomplete when taken at any
single instant. The same conclusion applies to space. If matter resolves
itself into vibration, then it cannot be located on determinate spatial
coordinates.

Quantum theory conceives the primordial elements of matter as the
patterned vibratory ebbs and flows of an underlying activity (see Bohm
2002: 12–13). There is a period that defines each system. Within that
period the system of vibration will oscillate between two stationary limits. This kind of system, constitutive of what we take to be a material
entity like an electron, 'requires its whole period in which to manifest
itself' (Whitehead 1967: 46). The question of where the material entity
is at any one instant can be answered only in terms of its average position
at the centre of each period. To divide time any smaller is to destroy the
coherence of the entity. Likewise, 'the path in space of such a vibratory

entity – where the entity is *constituted* by the vibrations – must be represented by a series of detached positions' (1967: 46). An entity is a process of traversing a series of positions at some frequency of oscillation. Process, rigorously conceived, is non-location. Material bodies do not resolve themselves at bottom into sets of minima located in space and ordered through time. Space and time are 'not productive of the ordered world', as they are for Kant; they are 'derivative from it' (Whitehead 1978: 72).

Localisability remains a requisite feature of matter mechanistically conceived and scientifically determined. Bergson follows Whitehead in denying simple location, and his too is an account of matter irreconcilable with the homogeneously extended bodies described by Newtonian mechanics. Bergson agrees that it is the Newtonian conception that is a scientific fiction and not a metaphysical fact. The inverse assumption – that simple locations ground dynamic processes – exemplifies what Whitehead calls (invoking Bergson's critique of spatialisation) the 'fallacy of misplaced concreteness' (1967: 64). This fallacy consists in taking for concrete what is abstract. Since abstractions are derived from concrete phenomena, the fallacy involves a mistaken inversion of the order of priority between the two. If localised material parts are an instance of the fallacy of simple location, then their postulation beneath the derivation of energetic material processes is an instance of the fallacy of misplaced concreteness. Matter so conceived is an abstraction. It is derivative on vibratory activity. 'The more physics advances', as Bergson observes, 'the more it effaces the individuality of bodies and even of the particles into which the scientific imagination began by composing them: bodies and corpuscles tend to dissolve into a universal interaction' (CE 188).

Thus, 'matter *extends* itself in space without being absolutely *extended* therein' (CE 203). Matter is defined by 'a tendency to spatiality', which means that 'although matter stretches itself out in the direction of space, it does not completely attain it' (CE 189, 207; cf. Worms 2004: 88–93). In 'conferring on matter the properties of pure space, we are transporting ourselves to the terminal point of the movement of which matter simply indicates the direction' (CE 203). So 'pure space is only the schema of the limit at which this movement would end' (CE 202). This schema is only ever incompletely realised in the material systems from which it is abstracted. The material systems 'are not in it', as Bergson puts the point; 'it is space which is in them' (CM 77).

Bergson not only appreciates the misconceptions about matter that Whitehead designates as fallacies. Bergson encodes Whitehead's fallacies into the constitution of matter itself, as matter's characteristic tendencies.

It is neither a real property (or state) of matter nor an epistemic error that lies at the basis of the fallacy of simple location. Matter tends towards isolability, mereological externality, and spatial localisability without ever being fully isolated, externalised, or localised. The fallacy of simple location is effectuated by (1) the intellectual elaboration of the concrete tendency towards spatiality to its terminus in the abstract geometry of homogeneous space, and (2) the extrication of that terminal pole from the tendency that it orients and its hypostatisation in the form of an independent state. The prosecution of the fallacy of misplaced concreteness involves the addition of another operation, (3) the retrospective postulation of the hypostatised state prior to the tendency towards it. The abstract, hypostatised state is taken for the concrete, the material tendency towards that state.

The fallacy of simple location elaborates itself through that of misplaced concreteness as a result of these three operations. The fallacy of simple location involves the extraction of a tendency internal to matter and its reification in the form of a state. The fallacy of misplaced concreteness involves the relocation of that state prior to the concrete material tendency towards it. Bergson's 'inert or unorganised matter' is just this sort of abstract conception, the terminal pole of a tendency reified as a state. It is Whitehead's fallacy of simple location rewritten as tendency internal to the matter it misconceives.

The material tendency towards spatialisation, understood as the tendency towards isolability, mereological externalisation, and the negation of the reality of time, is a tendency towards repetition. Only systems whose operations repeat in accordance with a set of laws can be isolated from external factors, decomposed into a set of constituent parts, and modelled mechanistically. As the effort to introduce as much indetermination into matter as possible, life is a tendency oriented towards the liberation of material systems from the law-bound necessity that characterises them. If necessity means repetition, then indetermination means novelty, that is, time properly conceived (CE 96, 4–7).

Tendencies are defined not only by the ends towards which they are oriented, but equally by the tendencies that oppose and inhibit them, constraining their ability to realise and resolve themselves. The tendency towards space requires an opposing tendency of this sort. This is the tendency towards time. In the next section, I map the details of this tendency through the inverse of the three facets of spatialisation. If the elaboration of the tendency towards space to its terminus furnishes the principle of inert matter, then the principle of life – that for which the *élan vital* is an image – should be understood in terms of the elaboration of the tendency towards time to its terminus too.

Temporalisation

Bergson defines life as 'a tendency to act on inert matter', the particular effect of which is unforeseeable, but 'always presents, to some extent, the character of contingency; it implies at least a rudiment of choice' (CE 96). The tendency to act on inert matter consists in 'the effort to engraft on to the necessity of physical forces the largest possible amount of *indetermination*' (CE 114). Matter is amenable to the necessity of physical forces only by being spatially stratified, and space is defined by isolation, externalisation, and localisation. Tending away from determinism, organised material systems are characterised by the tendency away from space. The tendency away from space is a tendency towards time. The two are directional contraries. The tendency towards time is what characterises organised material systems, or biological individuals. I define this tendency in terms of (1) the thermodynamic conditions and properties of organisation as the form of individuality proper to living systems, (2) the interpenetration of parts in the developmental whole of the organism, and (3) the non-locality of organisation, that is, its duration.

a. Individuation

The first difference between organised and unorganised material systems resides in their respective principles of individuation. The individuation of unorganised matter into articulated systems depends on a principle external to them, the perception of other living things or the spatialising operations of scientific study. Organised material systems or biological individuals are individuated through a principle internal to the processes that constitute them. Living things individuate themselves; they self-organise (see Miquel and Hwang 2016). Bergson situates this form of individuating activity within the paradigm of thermodynamics.

Bergson treats thermodynamics only briefly, but his discussion of its principles is significant (see Gunter 1991, Miquel 2010: 203–11, and DiFrisco 2015). The first law of thermodynamics, the law of conservation of energy, states that energy cannot be created or destroyed in an isolated system; it can only be converted (from potential to kinetic, for instance). Bergson considers this law 'relative, in part, to our methods of measurement' (CE 241–2). Since the law involves the quantification of energy within an artificially isolated system, it cannot be promoted to metaphysical status. The second law, 'the most metaphysical of the laws of physics', states that the entropy of any isolated system always increases (although Bergson himself never formally defines it). Entropy is a negative quantitative measure representing the degree of *dis*order, or the *un*availability of energy for work, in an isolated system. The higher its entropy, the more diffuse the energy in a system is and the less work can be done with it. Entropy is dynamic, as the energy in an isolated system is being continuously redistributed among all possible distributions of molecular collisions. The second law states that this redistribution tends towards equilibrium, a point at which there is no energy available for work as it has been diffused across all possible distributions and the system is completely disordered. This state is represented by the system's maximum entropy value. Understanding energy in terms of temperature, Clausius first recognised the second law as evidenced by the fact that heat spontaneously flows from a hot to a cooler body, but not the reverse (see Zemansky 1968: 215).

Bergson discerns in the second law a metaphysical insight, which he detaches from the form of a calculable magnitude in order to arrive at its definition as a tendency. Thus conceived, this law 'points out without interposed symbols, without artificial devices of measurements, the direction in which the world is going' (CE 243). This direction appears in the fact that 'all physical changes have a tendency to be degraded into heat, and that heat tends to be distributed among bodies in a uniform manner' (CE 243). Stated metaphysically, this means that

> changes that are visible and heterogeneous will be more and more diluted into changes that are invisible and homogeneous, and that the *instability* to which we owe the richness and variety of the changes taking place in our solar system will gradually give way to the relative *stability* of elementary vibrations continually and perpetually repeated. (CE 243, emphasis mine)

The progressive equilibration of potential energy can be extrapolated to the solar system as a whole, which is seen then 'to be ever exhausting something of the mutability [or availability of energy for work] it contains' (CE 243). Thus, the world seems to be 'unmaking' itself, diffusing the free energy required for the production of structure and complexity into heat, steadily tending towards the terminus of a cosmic equilibrium, or state of maximum entropy, from which nothing ordered can arise.

Bergson was born the year Darwin's *Origin of Species* was published, when the laws of thermodynamics had begun to confer the sort of unity upon the physical sciences that the theory of evolution would come to confer upon the biological (see Carvalho and Neves 2010: 365). These two systems are occasionally thought to present a contradiction. Roger Caillois is known for having proclaimed that 'Clausius and Darwin cannot both be right' (1976). Evolutionary theory seems to testify to a natural increase in complexity and organisation over time, while the second law of thermodynamics describes a universal tendency towards degradation and disorder. Among the first attempts to reconcile the order and complexity of living systems with the second law's prescription for entropic dissipation was Schrödinger's 'What is Life?' in 1944. His innovation consisted in the suggestion that living things maintain their internal order by absorbing free energy from their environment (in the form of nutrition, say) and 'exporting' entropy back into it by dissipating excess energy in the form of heat, sweat, and other kinds of waste (see Yoxen 1979). By modelling an 'isolated system' as a subsystem (the organism) coupled with its environment, the measure of entropy in the system as a whole would still increase in accordance with the second law, even while being resisted and reversed within one of its regions (the organism subsystem).

Around the same time, Bertalanffy defined the organism as an *open* system, maintaining 'itself in a continuous inflow and outflow, a building and breaking down of components. The organism, so long as it is alive, is never in a state of chemical and thermodynamic equilibrium; it rather maintains itself in a so-called steady state distinct from the latter (1968: 39; cf. 1932: 116 and 1950: 23–9). Steady states denote dynamic stability, sustained through matter-energy exchanges between a system and its environment. Where stability qua equilibrium in an isolated system denotes the unavailability of energy for work, steady state qua dynamic equilibrium in an open system denotes the relative constancy of each of the system's variables as an effect of patterns and cycles of interaction among parts, processes, and environment. The essence of life on this view consists in metabolism, the catabolic and anabolic processes of breaking down and building up organic molecules. Steady state stability in open systems is an ongoing metabolic accomplishment (see Kreps 2015: 134–7).

The most recent phase in the reconciliation between thermodynamics and sustained vital activity arrived with Ilya Prigogine's research on 'dissipative structures', a phylum of which the living system is a class. Prigogine argues that the physical sciences abstract time from their objects of study. His interest in thermodynamics stems from the search for a scientific ratification of the reality of time in the form of irreversible process (Prigogine and Stengers 1985: 10). Prigogine and Stengers appreciated Bergson's views on the positive efficacy of duration and noted that 'for Bergson, all the limitations of scientific rationality can be reduced to a single and decisive one: it is incapable of understanding *duration* since it reduces time to a sequence of instantaneous states linked by a deterministic law' (1985: 92; cf. Miquel 2010: 218–19).

Whereas classical thermodynamics conceived dissipation as internal to the tendency towards equilibrium, Prigogine saw in it a potential source of higher order. What he called 'dissipative structures', as for instance the Bérnard cell or the vortex, emerge from lower-level thermodynamic processes as forms of supramolecular organisation. The flow of energy and its dissipation through the system is what organises the system at higher orders of complexity. Order emerges in that it cannot be derived from molecular interactions. Its structure is imputed, instead, from the energy flow itself. The energy flow is what keeps the open system away from equilibrium with its environment. It is by remaining away from equilibrium that the dissipative structure maintains its organisation. The organism, on this picture, is a dissipative structure whose internal order is derived from the matter and energy flowing through it. In this way, life represents a temporary diversion of entropic decline.

Bergson's *élan vital* is a tendency whose principle cuts against the grain of the second law in accounting for the production of ordered complexity and the regional retardation of entropy. While it would go too far to suggest that he anticipated the developments in nonequilibrium thermodynamics that might be said to ratify his postulation of vital tendency (at least in part), the parallels are striking. Bergson insists, for instance, that there is to be discerned in living things, even if only locally and momentarily, 'an effort to re-mount the incline', that is, to produce – instead of to diffuse – structure, order, and complexity over time (CE 245). He concludes that 'everything happens as if [life] were doing its utmost to set itself free from these laws [of the increase of entropy]' (CE 245). He does not attribute to life 'the power to reverse the direction of physical changes', of energy into heat; he claims only that vital tendency acts or behaves 'as a force would behave which, left to itself, would work in the inverse direction', that is, upward against the degradational decline (CE 245–6).

Of course life is *not* left to itself; organisation occurs within the physical world that is governed by entropy. Vital activity cannot stop or reverse entropic dissipation. Bergson holds that it succeeds in '*retarding*' it (CE 246). Living things represent local and temporary 'resistances' to the universal decline. Organisms divert and delay the process of their own entropic degradation by exporting it outside of themselves. They do not succeed in reversing or even reducing the production of entropy in general. By 'retarding' it in themselves, they actually accelerate it elsewhere. This is the sense behind Bergson's otherwise difficult contention that living systems 'make' themselves on an organic time scale within a world that is 'unmaking' itself on a cosmic one (CE 249; cf. Miquel 2010: 230–47).

The explanation for how life 'makes itself', regionally delaying the global tendency towards 'unmaking', is a thermodynamic one. It has two components: '(1) a gradual accumulation of energy; (2) an elastic canalization of this energy in variable and indeterminable directions' (CE 255). By 'accumulation of energy' Bergson means negative entropy – what Schrödinger called 'negentropy' – that is, the potential for a system to do work (act, move, grow, self-regulate, and so on). Excepting the chemolithotrophs that feed on deep ocean vents, the fundamental source of energy for life on earth comes from the sun and is fixed in plants through photosynthesis in mineral form (CE 106, 115). Animals derive their energy either from plants, or from other animals that have derived their energy from plants, and distribute it across their internal systems in the form of proteins, carbohydrates, and fats, converted into different forms and allocated according to physiological needs (CE 116–17, 121). By the 'elastic canalization' of this energy, Bergson means its consolidation and preparation for use as potential, as well as its conversion, ordering, storage, and eventual release (and dissipation) in the performance of variable forms of work. That the canalisation of energy is 'elastic' means that it is plastic, responsive, adaptive to an array of organic needs and aims, from tissue repair to the regulation of internal temperature to locomotion and so on.

The wider a system's range of possible uses for stored energy, the higher the measure of its 'indetermination' as the index of what it can do. This is part of the essence of life, as 'the effort to engraft on to the necessity of physical forces the largest possible amount of *indetermination*' (CE 114). This effort manifests itself in the organised body's capacity to transcend its environmental constraints, realising potentialities that are not transmitted across it from the outside but that are internally produced.

To be free, classically, is to be the cause of one's own actions. Freedom finds its genesis in the introduction of contingency into a casually regulated material system. The kind of system capable of contingent, undetermined activity is already detached from external determination. To be so detached is to be organised. To be organised is to play a causal role in the determination of one's own activity. The nervous system is a faculty of hesitation. It receives stimulation, transmits it to relevant motor centres, and presents the living being with the largest possible number of responses to it (MM 20–1). The more possibilities there are for action, the less immediate the execution of any one of them is and the more the living being can hesitate before reacting. Hesitation operates in *Matter and Memory* analogously to the way indetermination does in *Creative Evolution*. The canalisation of energy and its preparation as potential for the performance of work conditions the possibility of self-determination in both senses.

Wherever the two conditions of accumulation and canalisation obtain, the global tendency towards entropic degradation is regionally delayed and life exists (see Miquel 2011: ch. 7). 'The truth', Bergson says, 'is that life is possible wherever energy descends the incline indicated by Carnot's law and where a cause of inverse direction can retard the descent' (CE 256). Carnot's law, or theorem, states that the efficiency of a heat engine that works by transferring energy from a hot reservoir to a cold reservoir, converting some of that energy into mechanical work in the process, is a function of the difference in temperature between the two reservoirs. Bergson likely has in mind the fact that energy flows from the hot reservoir to the cold reservoir and not the other way around, which means that it is governed by entropy. So 'a cause of inverse direction' would involve the exportation of entropy that is the concomitant of the organisation of an open system. This is consonant with the thermodynamic understanding of life.

Bergson's 'cause of inverse direction' is the tendency towards time. Physically speaking, it is realised in the constitution of the nervous system – if not necessarily in the form or mechanism in which we currently know it, then at least functionally, in its effects (CE 255–6). The nervous system, as the faculty for uniting sensory input with the motor apparatus, is what allows for the 'utilization of energy for movements', serving as the means by which accumulated energy is released (CE 120). The release of energy is the aim of its accumulation. Bergson considers the nervous system 'a master that the body serves' (CE 123). Energy might be accumulated for the sake of its release by means of the nervous system, but its use also 'preserves, supports, and maintains the organism in which the nervous system is set and on which the nervous elements have to live' (CE 121). The extraction of nutrients from food serves to 'repair tissues', provide 'the animal with the heat necessary to render it as independent as possible from changes in external temperature', and fuel the growth and maintenance of the cellular constitution of the animal itself.

It is only excess energy, accumulated beyond what is required to 'preserve, support, and maintain' the organism, that can be stored up as potential energy to be converted into movement (CE 122). The essential destination of acquired energy is there, in its accumulation (as potential) and release (as kinetic). The nervous system is part of the more general activity of canalising and redirecting the energy acquired from outside the system in question. Canalisation occurs across a variable but not random set of directions, as the set comes to be organised through the nervous system. To be canalised and prepared for locomotion, some acquired energy has to be used up in fuelling, repairing, and maintaining the locomotive system. The animal converts some of the energy that it acquires from its environment into its organisation and dissipates the rest (CE 121). So long as the movement of dissipation is counteracted by a continuous in-flow of free energy being converted into structure, the entropy prescribed by the second law is diverted, and the organism persists in an organised state away from equilibrium.

Bergson's thermodynamic definition of materiality as 'a reality which is unmaking itself' operates according to the degradational inevitability of entropy, while the temporalising tendency, 'a reality which is making itself', does so by converting free energy into structure, locally resisting the equilibrating movement of dissipation (CE 245, 248). These respective 'realities' appear as 'things' or 'states' – as physical molecules or physiological systems, for instance – only when intellectually conceived or scientifically determined. In fact they are actions or processes, continuous and interpenetrating, distinct by virtue of the directions in which they are oriented. Bergson characterises these directions vertically by equating vitality with 'ascent', against the equation of materiality with 'descent' (CE 245, 247, 369). The latter, as one scholar explains, 'is the overwhelming tendency of the physical world: systems tend toward their lowest accessible state of potential energy, and correlations among parts of the system become increasingly randomized over time, "forgetting" the initial conditions' (DiFrisco 2015: 64). Conversely, the movement of 'ascent', whose defining tendency is that of time, operates by increasing the potential energy available within a system, keeping that system away from thermodynamic equilibrium.

Organisation is maintained as a steady state via the dynamic play between movements of ascent and descent. Organisation is the quality of individuality proper to open systems. It is what Bergson calls a '*modus vivendi*', an agreement between the conflicting tendencies towards temporality and spatiality, the effect of which is the dynamic stability of the organism as a dissipative structure (CE 250).

According to Bergson's critique of inner purposiveness, the temporalising tendency cannot be posited as absolutely internal to the systems that it qualifies. As we saw, those systems cannot be fully individuated – mereologically, immunologically, genetically, or otherwise. There are a number of reasons for why this is so. First, the vital tendency is an individuating agency only inasmuch as it defines the direction of a concrete tendency inhibited from realising its aim of individuality. It is in conflict with the tendency towards space. It is also implicated – as I discuss in the next chapter – within life's global evolutionary finalism, according to which individuals are derivative on the movement through which they are constituted. That the systems embodying this tendency never achieve complete individuality follows. Second, life is an individuating agency in part, but it is not an organising one. Organisation is an effect of the relationship between opposing tendencies – a '*modus vivendi*' – when that relationship is defined by the predominance of the vital tendency. Organisation has to be maintained processually, for it is not given all at once, not even in the form of a principle.

b. Interpenetration

Self-organisation, as the individuating operation of the temporalising tendency, is opposed to the tendency towards isolability that characterises unorganised matter. As isolability has its mereological correlate in the externality of parts, there must be a counterpart to externalisation internal to the tendency towards time as well. The interpenetration of parts in a developmental system is what plays that role. Costelloe (1913) defines spatialisation and temporalisation by the principles of 'independence' (or externalisation) and 'interpenetration'. I think these constitute the mereological dimensions of the two tendencies, but not their defining features.

I draw an account of interpenetration from Bergson's examination of the structure and function of the eye, paying particular attention to the schemata of manufacture and organisation. The difference marked by the distinction between structure and function is the difference, for Bergson, between complexity and simplicity. The structure of the visual apparatus is complex. The eye is functionally embedded in a set of dynamical networks of membranes, muscles, cells, fluids, chemicals, vessels and glands. Vision depends upon the recursive reception, focus, transmission, and conversion of light into electrical impulse. Bergson contends that so long as these mechanisms are integrated and coordinated, 'as soon as the eye opens, the visual act is effected'. While the structure of the organ is complex, 'vision [its function] is one simple fact' (CE 88).

One of the dominant views in the philosophy of biology is that function does not necessarily describe the way a structure behaves or acts, but rather prescribes the kind of activity of which the structure should be capable and in which it should be engaged. Attributions of function are normative. The eye, if structurally integrated *in the right way*, should be able to see. Bergson departs from that view. On the dominant view – 'selectionism' – while function involves the attribution of a kind of teleology (it specifies what the organ in question is *for*), this is a teleology to be located in the selection history responsible for the retention of that organ in the present organism. As one philosopher explains, '"what it is for" can be spelled out as "what it was selected for" (e.g. for which effect it was maintained in the process of natural selection)' (Wouters 2005: 124; cf. Paul 1988).

Bergson does not think that function attribution can be reduced to a way of speaking about selection history. He thinks that there is a real tendency towards the function of vision that is being realised across the structures capable of generating it. It is 'this contrast between the complexity of the organ and the unity of the function' that is supposed to indicate an inadequacy both with the priority accorded by mechanism to the structural complexity of the organ as a condition for the generation of its corresponding function, as well as with the priority accorded by finalism to the function as the end in service of which the parts of the organ have been assembled (CE 88–9).

The difference between the structural complexity of the visual apparatus and the functional simplicity of vision is more radical than it might seem, but it is also harder to extrapolate across other organs and functions. The claim that function is simple does not mean that it has few parts, few roles, few goals, or few means of realising them; it does not mean that it is easy to model, replicate, synthesise, or to outsource to other structures. Functional simplicity is different in kind from structural complexity. The latter can be mereologically decomposed; the former is supposed to be (in principle) without parts. Bergson's conception of simplicity is a quality of experience, not a measure of its composition. This is why he does not begin by predicating it of any organic function whatever. He restricts its application to the function of 'an organ like the eye' (CE 88). This function, vision, serves as the clearest case of an experiential fact – and it is, at least at first, just this experience that Bergson designates as simple. When he claims that the distinction between function and structure is the distinction between simplicity and complexity 'in an organ like the eye', it is the organs of perception or sensation – organs whose complex structures generate sensory capacities the contents of which we experience – that he has in mind (CE 88). It is the experience of seeing that is supposed to be a simple or unified act, not or at least not necessarily the function of sight.

Bergson might be reproached for failing to clearly distinguish between structure and function on the one hand and function and experience on the other. Not all functions can be experienced. A healthy heart performs the function of supplying blood to the various systems of the body (see Ariew et al. 2002). It is a structurally complex organ, but is its corresponding function a simple one? If it is, it cannot be on account of its experiential simplicity, for there is no subjective 'inside' to the function, the way there is an 'inside' to vision as the experience of seeing. As I show below, it is the idea of an 'inside' that Bergson develops.

His conception of simplicity involves three moves. First, an equivalence is held to obtain between the functional simplicity evidenced in the experiential act of vision and the reality of movement itself. Then the concept of simple, unified movement is determined in the biological form of organisational activity, or what is better known as developmental process. Last, the continuity of developmental processes is prioritised over the differentiated components of the structures that those processes organise. The mereological complexity of organic structure is a *product* of the unity of the *process* of development.

1. The equivalence between the experience of vision and movement recalls *Time and Free Will*'s examination of movement as it is experienced from within (in time) as opposed to how it appears from without (in space). Raising one's hand feels from within like 'a simple, indivisible act' while, 'perceived from without, it is the course of a certain curve, AB' (TFW 90). That curve can be divided into a number of positions, and 'the line itself might be defined as a certain mutual coordination of these positions' (TFW 90). The line can be conceived in terms of its complexity. But positions and their ordering are only ways of diagramming what is experienced as a simple act of movement (TFW 91). When Bergson continues that 'the movement' or 'mobility' is in its 'indivisible simplicity' 'reality itself', he is anticipating the more radical conclusions of 'The Perception of Change', a lecture which he delivered four years after *Creative Evolution* (CM 125). By the time of the later text, simplicity, indivisibility, and unity are not only qualities of movement according to the subjective position of its experience in time (as when I raise my hand), but the essential characteristics of reality itself.

The transition from vision to movement seems to take place on the basis of their analogous experiential presentations. The localisation of a movement across a series of points, the decomposition of a unified phenomenon into a set of component parts, or the partitioning of an indivisible process into a series of states, not only misconceives the feeling of its experience, but falsifies the simplicity of the mobile phenomenon in its very objectivity as well. This occurs in two seemingly contradictory ways at once. In the complexity of spatial locations, component parts, or fixed stages of development there is both 'more' and 'less' than there is in the simple phenomenon itself. There is more, because the postulation and arrangement of positions requires the intellectual act of spatialisation and the work of ordering, whereas the movement itself contains nothing of either. There is less, because no matter how many positions are enumerated (following Zeno), they will never fully exhaust the unity of the movement that traverses them (CE 91). There is a simplicity to movement that is both irreducible to, as well as over-explained by, the potentially infinite complexity of its spatialised frames and forms. And 'just so', Bergson contends, 'with the relation of the eye to vision' (CE 91).

At this stage we can infer that the simplicity of vision is like the simplicity of movement in three respects. It is experienced as a unified phenomenon; it cannot be adequately reconstructed on the basis of its division into parts; and if the analogy with movement holds, then vision is a unified and indivisible whole in its essential reality as well. If this last inference obtains, then there is in the complexity of the visual apparatus both more and less than there is in the simple act of vision. This is how I understand Bergson's claim that vision is the inner truth of the structure of the eye. It is a claim made on analogy with the way the unified feeling of raising one's hand is the inner truth of the curve that describes its trajectory through space. This suggestion is crucial for Bergson's account of the convergent evolution of the eye, which I discuss in the next chapter. I leave it to one side for now in order to proceed through the next step in my reconstruction of Bergson's account of biological simplicity, the determination of the unity of movement in the form of organisational activity.

2. Bergson positions his conception of biological organisation against the manufacture model that both mechanism and finalism assume (see Al-Saji 2010: 153–4). To manufacture is to assemble pieces in view of some end. The finished thing can be considered 'whole' in the sense that its constituent parts are brought together around the end they share in common. The process works 'from the periphery to the centre', by coordinating a disparate set of parts in accordance with a final cause (CE 92). The final cause is at the centre of the manufactured object, ideally structuring the relations obtaining between its components. Manufacture thus proceeds 'from the many to the one' (CE 92). But the unity that defines the manufactured object (the 'one') is only ideal. The parts of a manufactured object (the 'many') remain external to each other, separable and decomposable, even after they are arranged. The manufactured object reflects 'exactly the form of the work of manufacturing it' (CE 92). Each of its parts represents a part of the work that went into it, and the whole is reducible to the work responsible for the constitution of its parts.

The Central Dogma is a good example of the manufacture model in more recent biology. The idea is that biological life is something like the computational result of genetic instructions. DNA makes RNA, RNA makes protein, and proteins make the organism. For each biological trait, there is a gene that codes for its development. The organism is assembled on the basis of a set of genetic blueprints. The Central Dogma assumes that the organismal whole can be decomposed into the parts that constitute it. Those parts can be resolved into genetic information in turn, which is given at the outset of development. The unity of the whole is ideal, closer to a building plan than a quality of organisation. That plan, realised through processes of transcription and translation, is what explains the structural integration of living things.

The scheme of organisation is supposed to differ from the manufacture model in each of these characteristics (see Lewontin 2000: 5). The organismal whole, understood on the schema of organisation, cannot be mereologically decomposed because the activity that organised it did not proceed, through assembly, towards the ideal integration of a diversity of disconnected parts (from the concrete many to the abstract one). Processes of organisation originate in a unity and proceed through the multiplication of parts, or the progressive differentiation and complexification of an originally indivisible simplicity. The parts of a living system are products of the developmental processes through which they were formed. Their analytic externality from each other is derivative on the developmental unity that underlies them. While the activity responsible for the organisation of the whole corresponds to the mereological complexity of the whole as a qualitative multiplicity, no such correspondence obtains between each part taken as a separate element and the developmental activity responsible for its formation. Development is continuous and qualitative. Separable parts are its abstract eventualities, not its building blocks.

Bergson invokes here 'the development of the embryo' only in passing, and while it is Raymond Ruyer – and later Deleuze – who will promote embryogenesis to the status of a general model for becoming, there is already operative behind Bergson's understanding of life as 'true continuity, real mobility, reciprocal penetration' an embryogenetic conception of change (CE 89, 163, 15; cf. Roffe 2019: 48–53). Bergson defines 'true continuity' as a process composed of a 'multiplicity of elements' characterised by 'the interpenetration of all by all' (CE 162). Interpenetration is a feature of the dependence relation between the qualitative character of the elements of a process and the whole rest of the process from which they are derived. To isolate a set of elements from the whole is to falsify them in the same gesture that constitutes them. It is to abstract from them the condition for their having become – and therefore for their being – what they are.

The development of the embryo evinces just this sort of 'true continuity' in that it cannot be determinately resolved into a series of independent elements. 'A fertilized egg', as Ruyer explains, 'is not a mosaic of territories that are irrevocably destined to engender this or that organ' (2016: 49). Early embryologists – Driesch and Spemann – found that if they grafted cells from one place to another on the embryo at an early enough stage of its development those cells would take on functions appropriate to their new locations. Grafted at later stages the cells developed as if they were still in their original positions. Development – they concluded – must proceed as if from the abstract to the concrete, from the more to the less indeterminate, from the equipotential (what can still become otherwise) to the fully actual (what is what it is). This process moves through three stages: presumption, determination, and differentiation. Before it is 'determined' as such, a given zone of the young embryo can only be 'presumed' to become some determinate structure. It can still become otherwise. Determination means that prior to the actualisation of some particular element, the zone that will become it cannot become otherwise, even if grafted elsewhere.

There occurs between equipotentiality and differentiation a phase during which each zone of the developing embryo is specified in terms of what it will become. Only after that determination does it begin to manifest the differentiated characteristics of a particular organic structure. What is most important here is that during the 'presumptive' stage of development, the embryo is equipotential, that is, its cells can be rearranged and their functional ends modified. Embryogenesis is a process through which equipotentiality is progressively minimised in a cascade of increasingly specific determinations: from the almost completely indeterminate, to some abstract theme (say, foot), to the particularisation of that theme (left foot), to a spatiotemporally specific, fully differentiated organic structure (*this* left foot). As the embryo is an 'equipotential territory' that is *not yet* what it will become, it can be mereologically decomposed into its component elements only retrospectively, after those elements have been formed over the course of their development.

When Bergson claims that the parts of a whole are interpenetrating, he does not mean that they cannot be abstracted and analysed independently. Organised wholes and continuous processes do admit of mereological decomposition, as we can posit in them whatever limits we see fit according to the dictates of some research programme and treat what is framed within those limits as an element (a state or part) of the whole. Thus different organic systems (circulatory, nervous, lymphatic) are abstracted from out of their implication in the organism and studied independently from each other. We can do the same with development, partitioning it into a series of states, each of which is changeless in itself, and conceiving the whole process in terms of its transitions from one state to the next. But while partitioning may be a necessary condition for analysis, it falsifies the dependence relation between each partitioned part and the qualitative whole from which it was abstracted. The reality of these parts is an effect of the partitions and spatialisations that condition them, and so cannot license the reduction of the whole down to them any more than they can be taken as reason to conceive the reality of change in terms of the states through which it is, so partitioned, said to pass. 'To say that the continuum [or whole] is composed of "parts" will therefore be false', as an early commentator explains, 'because parts are stable and there is no stability in a continuum' (Costelloe 1913: 146).

Here is an early anticipation of Developmental Systems Theory (DST). In their introduction to DST, Oyama, Griffiths, and Gray position the theory against the 'interactionist consensus', which consists in the claim that genes assume their causal powers within a set of internal and external environments that also help orchestrate the development of the whole organism. In contrast, DST understands causal attribution to be a matter of pragmatics: 'different groupings of developmental factors are valuable when addressing different questions' (Oyama et al. 2001: 2). The interactionist distinction between genes and any other causal factor in development, environment included, 'is just one more grouping, possibly helpful for some purposes, much less so for many others' (2001: 2). In her *Ontogeny of Information*, Oyama argues that chromosomal 'programs' for development emerge from among a set of inter-cellular relations developmentally structured in concert with factors outside the embryo (2000: 4, 13–14). The idea that some special causal nexus, like the genetic, initiates and presides over development makes sense only on the basis of an artificial partitioning of a process that precedes and exceeds it. The point, for DST, is that separable causal elements are abstractions. This is just what Bergson means by his claim that organisation is an affair of true continuity.

3. Once mobility is determined in its biological form as developmental process, the continuity of the process is prioritised over the parts organised through it. 'Life does not proceed by the association and addition of elements [as manufacture does], but by dissociation and division' (CE 89). Whereas manufacture integrates a set of distinct elements in view of a common end, organisation differentiates an initially indistinct multiplicity *away* from a common origin. Organisation is not only a matter of continuity, but a matter of 'real mobility' (CE 163). As with the fertilised egg, determinate parts come to be only over the course of development, as initially equipotential territories are progressively formed through differentiation. If organic structures are not assembled out of an association of elements it is because those elements are not given in advance. They are developmentally constituted along with the formation of the organs of which they are the parts.

Bergson says that the activity through which the living being is organised originates in a unified centre, an equipotential territory, 'and spreads around this point by concentric waves which go on enlarging' (CE 92).

By these 'waves' he seems to have had in mind what Spemann and Mangold (2001) would later call 'induction'. Their experiments confirmed that once a territory of the developing embryo was determined in view of what it would actually become, its effects would emanate outward, 'inducing' effects in surrounding territories and determining the developmental fate of the embryo concentrically. Spemann and Mangold called this initial determination an 'organization center', concluding that embryogenesis begins in 'a region of the embryo that has preceded the other parts in determination [the organization center] and thereupon emanates determination effects of a certain quantity in certain directions', that is, through induction (2001: 16). Opposite manufacture, organisation moves from the one towards the many. It passes from the virtual multiplicity of equipotentiality, through the elaboration of an increasingly distinct multiplicity of elements, towards the abstract pole of the isolable quantification of parts.

Interpenetration is not a property or state, but a tendency. It is the mereological facet of life imagined as *élan vital* and conceived as the tendency towards time. The perfect realisation of interpenetration would mean an absolute limit of indifference with respect to which no mereological distinctions could be made. This limit is approximated in the equipotentiality of the young embryo. Over the course of its development, embryonic tissue is progressively determined and differentiated, and its resultant elements are gradually externalised from out of each other. Interpenetration does not lure developmental processes in the form of a final cause into which they might resolve themselves, but rather insists through them as the unity of their common origin. This is an essential facet of Bergson's concept of duration: continuous processes of change retain their pasts. As I mentioned in Chapter 1, the structure of an organ is its very history of development consolidated. Differentiated organic parts, arising out of the initial unity of the embryo, retain and preserve that unity as they develop. The extent to which they retain their unified history corresponds to the predominance of the mereological facet of the temporalising tendency within them.

At the outset of development, distinct parts do not exist. However, that their unified origin insists through them as they are progressively differentiated means that those parts can be retrospectively attributed, in the state of interpenetration, to the developmental unity from out of which they were formed (TS 293). It is because the parts of an organic being retain the developmental history from out of which they were externalised that they can be subsequently conceptualised as having been internal to it (CE 23). Developmental interpenetration requires a partial act of retrospective attribution, just like the 'indistinct multiplicity' of elements in a qualitative heterogeneity. It is only after a set of elements are made distinct that they can be retrospectively understood to have been interpenetrating. The same should be said for the mereology of organisation (see Deleuze 2006: 95, 96, 100;). The analysis of the organism in terms of its parts regards it as a *product* – and not a *process* – of development, in abstraction from the organisational activity that generated it.

The process/product distinction corresponds to the relationship between the tendency towards temporality immanent to the organism as a developmental system, and the spatiality of the organism's material structure considered in abstraction from the process of its genesis. The material structure of the organ should be understood accordingly, as it is made tractable through decomposition for scientific determination, as a spatialisation of the unified temporal process through which it was constituted. The former is the product of the durational continuity of the latter, retaining and representing its developmental history in the spatialised freeze-frame of its finished parts. Its duration characterises the developmental processes responsible for the formation of the complex structure of the adult organism. But its duration does not terminate there, in the constitution of its fully formed parts from out of the state of their interpenetration in the unity of the whole. The duration of organisation continues through the ongoing activity on account of which the mature organism maintains itself as an organised material system. As I said, the organism is able to achieve a state of dynamic stability only by dissipating an influx of matter-energy. I turn now to the relationship between the organism's level of duration and that activity of self-organisation.

c. Duration

Bergson's nascent theory of metabolism may provide the key to a complete conception of the different levels of duration lived by different organisms. Organisms have to be grasped from within their delimited form as temporally open phenomena, streams of activity punctuated by the metabolic reproduction of their structures. Organisms experience their worlds in time in accordance with a rhythm of self-organisation.

Bergson refers to metabolism twice at the outset of *Creative Evolution*, in his early discussion of aging and its physicochemical explanation, and then again nearer the end of the text in his treatment of thermodynamics (CE 19–20, 34–5, and 251–5). The first occurs within a short discussion of aging as 'the insensible, infinitely graduated, continuance of the change of form' (CE 19). Mechanistic science stabilises and fixes the continuity of this change, partitioning it into a series of stages, states, or phases, each of which is treated as a relatively immobile portion of an otherwise continuous unfolding. The process of aging gets divided around the transitional periods that punctuate it – as in, say, puberty or menopause. These states are decomposed into the sets of physical elements that accompany and partially constitute them. Bergson refers to the elements that are taken to define senescence as 'phenomena of organic destruction' (CE 19). Mechanistic science is supposed to reduce the continuity of the process to these phenomena, as the manifest effects of aging. Bergson lists 'the facts of sclerosis, the gradual accumulation of residual substances, [and] the growing hypertrophy of the protoplasm of the cell' as examples (CE 19).

Bergson seems to have been ahead of his time in suggesting that the effects of senescence have to do with catabolism. The by-products of catabolic processes include lactic acid, acetic acid, carbon dioxide, ammonia, and urea. Their creation is usually due to an oxidation process involving a release of chemical free energy, some of which is lost as heat, the rest of which is used to drive the synthesis of adenosine triphosphate (ATP). Bergson suggests that the process of aging is linked to an increase in this kind of waste. Recent research suggests the same. One study argues that aging causes loss of many of the anabolic signals, while there seems to be a correlation between age and an increase in catabolic signals as well (see Roubenoff 2003). Another recent pair of studies argue for a conception of aging that locates its cause in 'catabolic malfunction', since catabolism helps facilitate a continuous turnover of damaged or obsolete biomolecules, and it is an insufficiency in this turnover, and a corresponding accumulation of 'biological waste material', that is manifest in the signs of aging (Terman and Brunk 2004 and Terman 2006).

By 'organic destruction', Bergson probably meant the degradational processes that convert organic compounds into inorganic molecules by breaking them down. He later substitutes the technical term of the time, 'katagenesis', for 'destruction' when referring to processes of the same kind, and credits Edward Cope's *Primary Factors of Evolution* (1896: 475–84) for having demonstrated his thesis: by restricting themselves to the study of elements of the 'katagenetic' order, physicochemical explanations of life deal only 'with the dead and not with the living' (CE 34–5).

These are antiquated spellings. The current spelling is 'catabolism', and variations are seldom utilised within biology. Cope's use of the term 'katabolism' came ten years after it was first coined in print by William H. Gaskell (1886: 46). Gaskell may have drawn upon an antecedent attributed by D. J. S. Doutrepont to his professor Dr. C. A. Wilmans 100 years prior (see Bing 1971: 170). It is not clear whether Bergson was familiar with Gaskell's work or the tradition that preceded him. It is possible that

Bergson took Cope to be the originator of 'katabolism'. Cope is known today for having proposed the law – 'Cope's Rule' – that animal groups exhibit a tendency towards the evolution of larger physical size in the absence of significant counteracting pressures or constraints. Metabolism, and metabolic needs, are thought to furnish one of the primary constraints on this trend (see Heim et al. 2015). Bergson finds in Cope the important distinction between two phases of metabolic process ('katagenesis' and 'anagenesis'), and the rule for which Cope would become famous has since been linked to the evolutionary significance of metabolism (see Kanamori 2010: 113–18 and Miquel 2010: 234–5).

Organic destruction – 'katagenesis' or 'catabolism' – is one of two subsets of a more general domain of organic activity called metabolism. Its other phase, which Bergson calls 'organic creation', or 'anagenesis', is now known as anabolism (CE 34). Catabolism and anabolism are interdependent processes that transform consumed organic substances, through biochemical reactions, into everything required by the organism for the maintenance of its life processes (see Dupré and O'Malley 2009: 2). Catabolism designates the conversion of complex organic compounds into inorganic molecules. This process frees the chemical energy stored in them, using it to drive ATP, and supplying the organism with a set of simple molecules that can be synthesised for other purposes (CE 120). These processes exhibit 'the fall, not the rise, of energy' (CE 35). This phrase presages Bergson's discussion of thermodynamics, in which matter is defined, as I mentioned, by the tendency towards equilibrium, which is characterised by 'the fall' of energy, its dissipation into heat and unavailability for work (CE 253–5). Catabolism is the metabolic correlate of the tendency towards materiality because the process through which organic compounds are converted into inorganic molecules dissipates the energy stored within them. Catabolism is the activity of turning life back into matter (CE 253). According to Bergson, 'it is only with these facts of the katagenetic order that physico-chemistry deals' (CE 35).

If catabolism breaks down, then anabolism builds back up. Anabolism is a kind of 'organic *creation*' (CE 20). Bergson associates it with the work of the '*vital impetus*' (CE 253–4). It is the storage of energy, and its conversion into structure, that allow the living being to resist the thermodynamic equilibration of energy in matter. Anabolism converts the molecules produced through catabolism into new organic compounds that can be used in the formation of cellular structures, the transmission of information, or as energy for the performance of other cellular functions. The role of the anabolic processes is 'to raise the inferior energies to their own level [that of the cell, or organism] by assimilating inorganic substances. They *construct* the tissues' (CE 34–5).

Bergson avoids speaking of metabolism as the name for the total set of these biochemical operations, preferring the terminology of creation and destruction instead, because he discerns within metabolic activity two directional contraries: the (vital) synthesis and (material) disintegration of organic compounds, the (vital) storage and the (material) release of energy. Anabolism creates organic compounds out of the inorganic molecules of catabolism, and catabolism disaggregates these compounds back into their constituent molecules in turn.

When Bergson insists that the processes through which organisms change and maintain their form cannot be exhausted through the study of 'the functional activity' of their metabolism, he is exploiting the directional distinction between anabolic and catabolic processes to furnish a second-order distinction between stability and change (CE 36). Metabolic activity consists in the interplay between the contrary tendencies of 'ascent' and 'descent', the rise and fall of energy, its capture and release (CE 14, 279, 295). Where tendencies interact, a certain rate or rhythm of vacillation occurs between them (CE 128–9). Earlier I mentioned that in the case of the perception of a sensible quality, like colour, the appearance of a fixed form or persistent identity is the effect of a rate of the vibration of elementary movements. Bergson returns to this insight late in *Creative Evolution*, concluding again that 'every quality is change', the appearance of a state being the effect of changes occurring at some rate (CE 301). He adds that the same point holds for the living being, the 'persistence' of which 'consists in a series of palpitations', analogous to the fact that 'the permanence of a sensible quality consists in this repetition of movements' (CE 301).

I think this suggestion should be replaced within the discussion of metabolism. In addition to the metabolic difference between anabolism and catabolism, Bergson invokes a second difference between the processes that make up the organism's metabolism and their chemical contents and products (CE 36). The physicochemical study of organisms is concerned with these contents and their production in metabolic cycles, whereas 'those whose attention is concentrated on the minute structure of living tissues, on their genesis and evolution, . . . are interested in the retort itself, not merely in its contents' (CE 36). A retort is a piece of glassware used for chemical distillation. Liquid is heated in its base, and the downward-pointing neck of the retort allows the vapour of the heated liquid to condense and flow into a collection vessel. The retort is a material form in which the process of distillation can take place. The 'retort' of metabolism can be thought to consist in the 'metabolic pathways' through which the linked series of catabolic and anabolic biochemical reactions are executed.

To pay attention to the 'retort' instead of only analysing its chemical contents is to 'find that this retort creates its own form through a *unique* series of acts' (CE 36). Metabolism should not be understood as a constituted system of pathways, but as a sequence of activities through which those pathways come to form themselves over time. The metabolic activity through which the organism maintains itself does not occur within an already-formed physiology, but rather comes to form that physiology over the course of its operation. In this respect, it is worth noting that 'metabolism' was coined as a noun after the establishment of the adjective 'metabolic', becoming a thing in the place of a process (see Landecker 2013; cf. Bing 1971: 166). The appearance of organic form can be conceived as an ongoing achievement of the open-ended reiteration of a series of metabolic processes. The organism appears to be stable in just the same way that a sensible quality appears to be static.

'Palpitation' is Bergson's term for vibration's organic analogue (CE 301). The term usually refers to the beating of the heart. Like anything with a rhythm, the heart oscillates between (at least) two states. Its cycle consists in the regular alternation between systole and diastole, a phase of contraction that empties the heart of blood and a phase of expansion that refills it. Heart rate is a measure of the frequency at which each state recurs in a cardiac cycle. This scheme is not particular to the heart and can be applied to any phenomenon that persists through rhythmic self-reiteration. Bergson uses the term palpitation one other time, in his 'Introduction to Metaphysics', in order to refer to the oscillating elements behind the appearance of any stable form whatsoever (CM 158).

Like the cardiac cycle, metabolism consists in the alternation between two phases of a process. Catabolism and anabolism are the systole and diastole of metabolic activity. The rhythm of their interplay is measured by what biologists call a 'metabolic rate' (see McNab 1997: 718). This rate expresses energy expenditure per unit of time by measuring the amount of heat generated by an animal body at rest. The easiest way to ascertain metabolic rate is by measuring 'basal' rate in endothermic animals, which was standardised for humans by Dubois in 1924 and applied across domesticated mammals by Kleiber and Benedict in the 1930s. Basal Metabolic Rate (BMR) 'is the rate (1) in the zone of thermoneutrality when the individuals are (2) inactive, (3) postabsorptive, (4) adult (thereby eliminating the cost of growth), (5) nonreproductive (eliminating the cost of pregnancy, lactation, egg formation, or incubation), and (6) regulating body temperature, and it is (7) measured during the inactive period' (McNab 1997: 718). BMR is measured in energy units per unit time ranging from watt (joule/second) to ml O2/min or joule/hour per kg body mass J/(h·kg).

Heat is generated by a resting body as an effect of what it requires to maintain itself in operation. In endothermic animals, heat is generated as a by-product of the biochemistry of circulation, regulation, repair, growth, contraction, and digestion. Higher rates of activity express cycles of anabolic and catabolic processes of uptake and dissipation that pulse at rapid, more energy-consumptive frequencies. Their rate can be read as a measure of the time or speed at which self-organisation operates.

If organic form is generated and maintained as a global effect of a series of metabolic processes, then it cannot be reduced to or explained by the (biochemical) elements in which those processes consist at any one moment in time (CE 36–7). Organic form exists as a global quality of the processes that sustain it. It is non-local to its elements in the same way that a melody is (MM 148, TFW 101). Form has to be grasped as a unity realised over time, implicated in each of its elements without being reducible to any of them independently.

Organic form also consists in the manner through which its constituent parts unfold over time. Form is the articulation of a stream of elements, the temporal organisation of their sequence in space. 'Form is essentially extended', as Bergson says, 'inseparable as it is from the extensity of the becoming which has materialized it in the course of its flow' (CE 318). Form is non-local to any one moment of its history (see Grosz 2007: 25 and Prochiantz 2008). The appearance of any one stable form that would evince an identity of the organism with itself is 'only a snapshot view of a transition' (CE 302). If form is understood in terms of stability, then it exists in abstraction from the continuous processes of individuation. 'There is no form since form is immobile and reality is movement' (CE302). What we discern as stable forms are just relatively slow periods of becoming.[1] What we discern as the transitions between them are just changes 'considerable enough to overcome the fortunate inertia of our perception' (CE 302). Beneath the thresholds and fusion rates of perception, 'the body is changing form at every moment' (CE 302).

Change is heterogeneous. 'Becoming is infinitely varied' (CE 304). Different processes of transition unfold at different speeds, as different vibratory phenomena oscillate at different frequencies. Objects appear stable and qualities appear distinct because they are changing at different rates and rhythms (CM 121). As I mentioned in Chapter 2, in the case of perceptual forms speed is always relative, which means that the appearance of formal stability is an effect of the difference between the rate at which the object or quality is changing and the rates and thresholds that govern the perceptual systems of the perceiving subject (CM 122). The moving and changing elements of sensible objects are contracted in the perceptual registration of stable and persistent qualities like colour (CE 301). Perceptual form is an achievement of the contraction, at some rate, of a continuous becoming. The intensity of the contraction is what determines the constitution of the form.

These contractions are regulated according to the degree of tension that defines a given consciousness (or set of sensorimotor apparatuses). Durations differ from each other across a spectrum of intensities. This spectrum can be arranged in the form of a hierarchy. At its bottom is matter, the rapid oscillations of which do not endure but repeat outside of each other (CM 158). At this level there is less intensity of duration than there is extensity of space. A consciousness capable of registering individual material repetitions would be so relaxed as to be diluted across extensity and in fact not a consciousness at all (see Dolbeault 2008). It would be dissociable and localisable across each of its moments. Unable to determine its own actions, it would be part of the material relay of causes and effects. Advancing up the hierarchy, 'we go toward a duration which stretches, tightens, and becomes more and more intensified' (CM 158). Higher levels of duration are marked by the contraction of more elements in each moment (MM 279). The intensity of a given consciousness of duration is a measure of how many differences are condensed and registered in one of its moments. Intensity is a measure of the synthesis of differences. The more contracted a level of duration is, the less it can be localised in any of its moments and elements.

More intense consciousnesses distinguish themselves further from outer durations by contracting more changes outside of them into denser, more actionable impressions. The consequence is a widened gap between subject and world, and a concomitant expansion of the array of actions and reactions of which the subject is capable. There is a relationship between tension of consciousness and ability to act. 'The greater the power of acting bestowed upon an animal', as Bergson explains, 'the more numerous, probably, are the elementary changes that its faculty of perceiving concentrates into one of its instants' (CE 301). The relationship between action and perception is manifest in the coordination of sensory and motor organs (CE 300). It expresses the fact that organisms perceive in order to act. 'The tension of their faculty of acting is', as Bergson puts it, 'probably proportional to the concentration of their faculty of perceiving' (CE 301, 188). Both action and perception can be conceived in intensities, measured by the tension of consciousness lived by each organism at some level of duration. Time is behind both perception and action alike.

Metabolism is what provides the ground for the unity of action and perception. As that unity is lived in time, the organism's metabolic rate is what grounds and explains its place in the hierarchy of durations. It is its metabolism, ultimately, that unifies the tension of consciousness of each organism with its perceptual fusion rates and ability to act. *Metabolism is the biological foundation of lived time.*[2]

The first step in the elaboration of this contention consists in a metaphysical reconception of metabolism as the inner principle, or 'inward articulation', of an organism's formal changes (CE 311). From the outside, organisms exist as indivisible processes of change (CE 312). This indivisibility, 'the insensible, infinitely graduated, continuance of the change form', is what is 'properly vital' in organised matter (CE 19). The apparently stable forms assumed over the course of that change are really only '*possible stops* imagined by us, from without, along the continuity of a process' (CE 312). There is no stable entity behind that continuity; it is rather the subject of itself (CE 313; cf. CM 125).

Change is its own subject in another sense as well. Bergson claims that 'all movement is articulated inwardly', and that there is an 'internal organization' to 'vital evolution' (CE 310–12). An organism's changing reality of form '*endures* inwardly', and it is in this interior dimension that the lived duration of the organism consists (CE 363). To say that change of form is the subject of itself is to say both that there is no stable entity of which it is predicated, and that it is lived from the inside in a manner particular to the organism changing.

Organisms do not live the entire history of their development as a single unified reality, as if at a stroke. They inhabit particular zones of relevant activity, defined as a present in relation to the history that they consolidate and the actionable future that is opened for them (CM 126). The tripartite structure of this time complex is the form in which the organism's outer becoming is inwardly articulated. This is how organisms apprehend themselves from within, in time (CE 363). The question of the duration of organised matter is the question of the relationship between a process of change and its inner articulation. It is the question of how or by what means a process of change comes to be apprehended in the form of a past-present-future complex.

Metabolism is what furnishes this means. Metabolism is comprised of the set of processes that determine the rate and rhythm of organisms as dynamic systems. Metabolic rate is an expression of the number, speed, and complexity of the elements being processed, synthesised, and dissipated through the catabolic and anabolic processes per unit of time. Rhythm is harder to codify. It refers to the interaction of the catabolic and anabolic processes in time-regulated cycles of oscillation (see Novak and Tyson 2008: 981). These oscillations space out each moment of the metabolic process from the moments subsequent and prospective to it.

Rhythm designates the relationship between each passing moment in a temporal sequence. It is a crucial facet of metabolism, as rate alone does not indicate how the metabolic processes unfold in a dynamic and interactive sequence. Rate measures the energy expended by the metabolic processes. Rhythm designates the character of how they expend that energy over time.

An organism might run a high metabolic rate because of a slow digestive system, a set of energy-intensive mechanisms for breakdown or dissipation, hyperactivity, stress, large body size, and any other number of reasons, without necessarily processing its world faster or more continuously. This organism would run a high metabolic rate at a slower relative rhythm. Smaller or more energy efficient organisms might run lower metabolic rates at faster rhythms of uptake and dissipation. The two dimensions have to be held together.

Rhythm and rate are like the form and content of self-organisation. Metabolic rhythm furnishes the structure or form of the past-present-future complex by determining the length and spread of a present in relation to the past that it follows, like a pulse, and the future into which it fades (MM 272). Rate is a measure of the elements or content contracted and dissipated in each passing pulse (MM 273). Organisms live their changes in contracted pulses or moments of time. The 'palpitations' that constitute the living consist in the punctuated reiteration of structure across the continuous becoming of organic form (CE 301). Life is a rhythmic phenomenon, and its rhythm is a matter of the oscillatory metabolic cycles of self-organisation (CE 128).

An organism's place in the hierarchy of durations is determined by the metabolic relationship between organising movements of ascent and equilibrating movements of descent. A predominance of the former manifests itself in the sustained uptake of free energy that allows the organism to maintain itself in dynamic stability far from the equilibrated energy of its environment (CE 252–3). It manifests itself in higher levels of organisation (CE 262). Higher levels of organisation ground higher degrees of durational contraction (CE 301). Metabolism is itself a contractile network of processes. The rate and rhythm of lived duration correspond to the rate and rhythm of metabolic contraction. The intensive degrees of the durational hierarchy are indexed to differing relationships of predominance and subordination among the catabolic and anabolic processes of different organic systems.

Bergson defines duration as (1) the retention of a past history that is (2) contracted in a temporally extended present, and (3) opening onto an unpredeterminable future. The metabolism of self-organisation is at the base of each facet. The retention of a past history of acts requires some means by which the series can be brought and held together through successive unfolding. Succession alone 'does not constitute time any more than it causes it to disappear', as Deleuze explains; 'it indicates only its constantly aborted moment of birth' (1994: 70). The constitution of a history requires, at minimum, the endurance of the entity whose history it is and in whom that history is retained. It requires that this entity persist across the series of acts that constitute it in order to retain them in the form of a past. Metabolism supplies this mechanism for organic persistence. It does so by 'creat[ing] its own form through a *unique* series of acts that really constitute a *history*' (CE 36). The metabolic processes operate in order to maintain an organic form that persists across them. They are retained in the form that they create, constituting a history for that form at the same time as they underwrite its persistence. To persist in form over a course of changes is to endure in time. To endure in time is to retain a history of past changes (CE 37).

By unifying a series of acts with the preservation of form across time, Bergson's conception of metabolism prefigures the autopoietic systems theory of the late twentieth century. I traced the outlines of this theory in Chapter 1. There I considered the idea that the structure and organisation of an organism is an achievement of its own reiterative activity. Autopoiesis, or self-production, refers to the network of processes that sustain the unity of the organism, which 'through their interactions and transformations continuously regenerate and realize the network of processes (relations) that produced them' (Maturana and Varela 1980: 77–8). Organisation consists in the repetitive regeneration of the organism itself over time. Persistence, or enduring unity, is an effect of a running series of acts. This is how Bergson understands metabolism, as that series of acts whose succession maintains an organised form in which they can continue to succeed each other.

Bergson is at odds with autopoiesis, however, in that form is for him ultimately nothing beyond the prolongation of each act in a history (CE 2). There is no real or emergent unity that those acts serve to maintain, no substance behind their changes. The appearance of form as a stable reality is due to a set of intellectual habits and perceptual thresholds. It is not an ontological achievement but a practical heuristic (CE 312). Organic unity consists in the persistence and continuity of the acts that constitute it and nothing besides. It *is* those acts, as they are prolonged through each other in time at some rhythm and rate. The duration and persistence of the organism are two dimensions of the same activity. Organic time is metabolic continuity grasped from within. *The retention of the past in the form of memory is the durational correlate of the metabolic activity of organisation.*

The second facet of duration is the contraction of a retained history of change in a temporally extended present (CE 4–5). This is partly what it means to retain a past. If every present were a distinct and autonomous unit of experience, then one would succeed the other without unity, and there would be no history available to them. The retention of the past implies that it bears some effective relationship to each present that passes (MM 191). The past is more or less effective in the present in proportion to the tension of the duration in which the two are united. Intensity is a measure of contraction, and the past is registered in and prolonged through the present as a correlate of how many changes are contracted in the present (MM 194). The same point holds for memory. Less intense consciousnesses have less of their past histories available to them; they react to present stimuli in more or less invariable ways, independently of how many times they have encountered the same impressions before (MM 92–3). More intense consciousnesses can arbitrate present decisions on the basis of past experience. Actions become freer in proportion to how much of the past can be leveraged for present purpose, prolonged through present action, and contracted in present perceptions. We should be able to see this tendency manifest, to some degree, wherever there is life. 'Wherever anything lives, there is, open somewhere, a register in which time is being inscribed' (CE 20). Every living thing is 'a thing that *endures*', which means that 'its past, in its entirety, is prolonged into its present, and abides there, actual and acting' (CE 15). How much of that past can be utilised is a measure of the intensity of the consciousness of the organism, but it is contracted there nonetheless.

Dennis Bray's work on *E. coli* seems to have borne some of this out. Bray demonstrates that in the nutrition-seeking behaviour of *E. coli* there is an effective bacterial memory. The bacteria 'continually reassess their situation' by means of a 'a sort of *short-term memory*' (2009: 7). *E. coli*'s memory is tested by measuring its response to an incremental adjustment in the concentration of an attractant like aspartate. *E. coli* responds to change. It stops responding once the concentration of attractant has settled into equilibrium. In Bray's estimation, 'by measuring the rate of change in the signal [the aspartate], the receptor cluster [the bacteria] has in effect performed calculus!' (2009: 94). The bacterium performs a differentiation, drawing a difference from its repeated measurements of the aspartate. It synthesises that difference in adapting to fluctuations in the concentration of the aspartate.

Navigating a field of concentrations of attractant, the bacterium preserves its past in the form of an adaptive pattern, and integrates the retention as it follows a projected trajectory. The interval between its past and future, between adaptive retention and projection, is approximately 10 seconds. In a similar study, Howard Berg writes that 'this [interval] sets an upper limit on the time available for a cell to decide whether life is getting better or worse. If it cannot decide within about 10 seconds, it is too late' (2004: 49–50). The cell must repeatedly draw differences from its environment if it is to navigate successfully. Every contraction is implicated in a relative duration, and every duration is rhythmically articulated.

Bacterial memory evinces the relationship between the prolongation of a retained past into a perceptual present and the metabolic rhythm and rate of self-organisation. Metabolic rhythm calibrates for a limit on the temporal spread of the perceptual present by articulating how much of the past can be contracted, while metabolic rate calibrates for a limit on how many changes can be registered in each passing present impression. Manuel DeLanda arrives at the same insight in the context of the relativity of temporal oscillation. 'What is', in his terms, 'immediate past and future for [a biological oscillator] would still be part of the "lived" present of an oscillator operating at longer time scales, at the level of geological or stellar dynamics, for example.' At the same time, this biological present 'already includes many past and future events for oscillators operating at atomic and sub-atomic scales' (2002: 88). Different systems contract presents at different rates. The present of one, if long enough, may include events already perceived as past by another.

These rates of synthesis vary according to oscillation scale. Biological phenomena operate at higher rates than geological rhythms. Within the biological register, rate is relative to the metabolism of the organism in question. The composition of its present, as well as the fusion rates that measure it, are a function of the rate and rhythm at which the organism self-organises through time. The organism lives its duration, in the present, in accordance with this rhythm. It is mediated from itself and the world by the metabolic rate at which its present is contracted. Organisms whose metabolic systems pulse at higher rates tend to perceive the world at higher frequencies.[3] 'Each level of temporal scale defines', for DeLanda, 'what oscillators at that level "perceive" as *relevant change*: certain cycles are simply *too slow* for them to appear as changing or moving relative to a faster level, and vice versa' (2002: 89–90). *The contraction of the past in the present is the durational correlate of the metabolism of the living being.*

The third facet of duration is that every present is opened onto an unpredeterminable future. As every present is a contraction of the past, and as every present is added to the past that will be re-contracted in a subsequent present, the relationship between the two is a dynamic one.

The past is not a static repository of passing presents but an evolving dimension of every present that passes. The future can be determined neither on the basis of the past that conditions it, nor as a causal forecast of the presents whose passing it succeeds. The future consists in the causal non-closure of the temporal complex. The causal relationship between the past and present is held open by the reconfiguration of the past through its incorporation of every new present in which it is contracted. The future is the temporal category of that openness. It designates the non-coincidence of present and past. The wider the non-coincidence, the more effective the future is and the harder it is to determine in advance. Duration's constitutive novelty is the measure of this indeterminability.

The relation between the uptake and dissipation of energy through the metabolic pathways is a relation obtaining between past and present. As the channels through which energy is dissipated multiply in number and become increasingly plastic in shape, similar kinds and amounts of captured energy become expendable in a widening array of directions. The more sophisticated a metabolic system is, the higher the inverse proportion between energy already captured and the channels through which it can be expended. The present of action is liberated, for complex metabolic systems, from the past of the energy dissipated through it.

Indetermination is a metabolic accomplishment. The tendency towards metabolism is synonymous with the idea of life itself, which 'seems in its essence like an effort to accumulate energy and then to let it flow into flexible channels, changeable in shape, at the end of which it will accomplish infinitely varied kinds of work' (CE 253–4). Duration's constitutive futurity consists, at bottom, in an appearance to consciousness of this primarily metabolic fact. *The unpredeterminable future of the living being is the durational correlate of its canalisation of energy.*

Conclusion

Duration is a function of metabolic self-maintenance. It is not the essence of life. Neither is interpenetration or self-organisation. Life has no essence. It is not a kind, category, property or feature. Life is Bergson's name for cases of the composition of spatialising and temporalising tendencies when the latter predominate over the former to some degree of intensity. Those are cases of composite material bodies able to maintain themselves in a dynamic stability away from the equilibrium of their environments by channelling and metabolising energy from outside. These cases are each organisms. Their life consists in the realisation of the temporalising tendency in the form of self-organising processes, the developmental interpenetration of their component parts, and the level of duration they live as a function of their metabolic canalisation of energy. In this way, living beings are furnished with their particular qualities of individuality, mereology, and subjectivity.

But life is not exhausted by the organisation of organic forms. Life is an event unfolding across two time scales simultaneously, the individual and the evolutionary. The *élan vital* is an image not only for the life of individual organisms, but for the continuity of the evolutionary process as a global fact. As an evolutionary tendency, life pursues a directional progress through the medium of individual organisms, tending towards indetermination not only in the material complexity and durational intensity of particular forms of life, but through them and by means of them. Individuality is nested within an evolutionary movement that unites all extant organisms by impelling them from behind.

5

Finalism Inverted

Individual living systems are the parts and passing moments of a global evolutionary movement. When evolution is considered as a single, still-unfolding event, the *élan vital* is an image for a unified tendency realised through and dissociated across the medium of particular organisms. The idea that evolution evidences a kind of directionality is called orthogenesis. The qualified finalism of Bergson's position holds that life is oriented from behind, by a shared impulsion, not from ahead, in the form of a pre-existent goal.

Introduction

Individual living systems are defined by the tendency towards temporalisation. They are also the medium through which a global tendency unfolds itself. Each organism is unified in the evolutionary movement. They are its parts and passing moments. The *élan vital* is an image for the tendency that unifies, patterns, and directs the evolution of life on earth. This is the idea behind what Bergson calls the 'psychological interpretation' of evolution.

This chapter begins by distinguishing between the individual and evolutionary registers on the basis of the time scales proper to each. The temporal difference between them corresponds to the conflict between the metabolic self-organisation of individuals and the reproductive continuity of their lineages. I specify Bergson's position on continuity by elaborating upon his engagement with Weismann's theory of the germ-plasm. Bergson retains from Weismann the idea of a line of variational energy that runs through individual organisms. Bergson holds that variations are internally impelled, and therefore directed at their source, shaped and diverted via adaptive pressures secondarily. This position is known as 'orthogenesis'. It provides the key to Bergson's reformulation of finalism, as a variant of the externalism that was introduced in Chapter 1. This form of finalism consists in the idea that life, as an evolutionary movement, is a virtual whole immanent but external to each of the individual organisms that instantiate it. Their various lineages are not oriented towards a pre-existent end but rather directed from behind by a common tendency. It is this common tendency that receives a 'psychological interpretation'.

This chapter has five parts. (1) 'Rhythm and reproduction' and (2) 'Weismann redux' extend the last chapter's analysis of organisation by distinguishing between the individual and evolutionary facets of life, first in terms of time scale, then in terms of a continuity of evolutionary development across the discontinuous individual instances of particular organisms. (3) 'Orthogenesis' isolates and elucidates Bergson's view on the directionality of the evolutionary movement. (4) 'Vitalism in question' distinguishes Bergson's account of the nature of life from the neo-vitalism with which it is typically associated and suggests that the category of finalism is more appropriate. Finally, (5) 'True finalism' elaborates Bergson's variation on finalism according to his 'psychological' conception of directionality.

Rhythm and reproduction

Life is a category that scales. It refers to the organisation of individual bodies as well as the historical processes of speciation through which populations evolve.[1] The temporality of life is striated across its individual and evolutionary registers. It has a direction particular to each. The directional conflict is clearest in the relationship between what textbook biology considers the two central features of life: metabolism and reproduction. Metabolism manifests the tendency towards individuality, as it is through the conversion of energy into structure that organisms are able to persist. But 'while the tendency to individuate is everywhere present in the organized world, it is everywhere opposed by the tendency towards reproduction' (CE 13). In Chapter 1, I mentioned that reproduction requires the continuity of a set of components across at least two generations of individuals. This continuity opens what would otherwise be the self-contained individuality of the organism downward onto its descendants and upward onto its progenitors. 'Its very need of perpetuating itself in time condemns it never to be complete in space' (CE 13). The individual is partly a derivation from the genetic line running through it. It is the genetic line that selection acts upon, the lineage that evolves. The tendency towards reproduction manifests the tendency towards evolution through speciation. Life unfolds in a play between individuality and evolution, realised through the metabolic and reproductive processes; and 'the biologist must take due account of both tendencies in every instance' (CE 13).

This remains an important admonition. At least since the mid-nineteenth century, metabolism and reproduction have been understood as two equally important but biologically distinct characteristics of life (see Landecker 2013). They are relegated to separate temporal organisations. Metabolism has to do with the persistence of the organism throughout its life span, while reproduction has to do with the persistence of biological substance across subsequent generations. At the end of the nineteenth century, this distinction was mapped onto the time scales of the soma and germ, and the current disciplinary divisions between molecular biology and biochemistry maintain the separation (see Gilbert 1982: 157–9). Over the course of the twentieth century, with the rise of molecular biology and genetics, metabolism has become increasingly subordinate to reproduction as a defining feature of life. As molecular biology came to determine the facts of reproduction and heredity through the study of the DNA molecules that constitute the genome, the biochemical processes of metabolism came to be understood accordingly, in the form of a 'program' 'run' by the genetic code.

This hierarchical division of metabolism from reproduction is currently in the process of revision. By demonstrating the relevance of metabolic facts for genetic (and epigenetic) processes, recent research has begun to destabilise the historical and disciplinary separation between metabolism and reproduction. Here is one example. High-fat diets in male mice leave an imprint on the metabolisms of their offspring, even in the absence of any physical contact between parent and child, likely through the dietary alteration of the epigenetic molecules such as microRNAs that are transmitted through sperm (see Ng et al. 2010). Research of this sort has prompted an effort on the part of philosophers of biology to reconceive the roles played by metabolism and reproduction in a more nuanced conception of life. Some argue that life is something that arises only through the complex interaction of these two features. Evolution should be understood to act on reproductive-metabolic systems instead of 'traditional' individual organisms (see Dupré and O'Malley 2009). Metabolism is something that happens in collaboration. Humans, for example, do not digest and metabolise as individuals, but in symbiotic concert with the bacteria of their microbiome. Complex multicellular organisms are incomprehensible as living systems apart from their microbial alliances. Philosophers of biology Dupré and O'Malley conclude with proponents of the hologenome concept of evolution, such as the Rosenbergs and Roughgarden et al., that 'the unit of selection, the entity in which selfishness may perhaps be expected as the norm, is a collaboration of many different lineage-forming entities' (2009: 14; cf. Roughgarden et al. 2018). Living systems are metabolic wholes composed of the association of many organisms of different lineages. The collaborative whole serves as the subject of reproduction and evolution. The two categories are, if conceptually separable, biologically inextricable.

Bergson has been positioned in the history of metabolism as an early critic of the separation of metabolism from reproduction (Landecker 2013). On this account, Bergson is noteworthy for having insisted that biologists take due account of both, but he did not go far enough in complicating the distinction between them. Metabolism and reproduction are testament, after all, to conflicting tendencies. Both tendencies are operative in individual organisms, which is why they should be studied in conjunction, but they are in principle separable and distinct nonetheless. That is partly true. However, by conceiving metabolism and reproduction as tendencies Bergson was also able to effectuate (and so anticipate) their integration. Bergson did not understand metabolism and reproduction as separate dynamics interacting in individual organisms, but as continuous and mutually presupposing tendencies. Metabolism and reproduction are indissociable both in principle and in fact.

Bergson's difference between metabolism and reproduction appears at first to be a difference in direction (CE 13). Wherever there exists an interplay between directionally contrary tendencies, there exists a rhythm that defines their relationship (CE 128). The integration of metabolism with reproduction is a matter of this sort of rhythm, occurring between the organism as a persistent metabolic system and the organism as a 'place of passage', or a moment in a lineage that precedes and succeeds it (CE 128). This rhythm differs from more familiar rhythms of oscillation – such as between the anabolic and catabolic processes – in that the tendencies towards metabolism and reproduction are not in principle antagonistic to each other. It is only in fact that they appear to be. They are not in principle antagonistic because they are defined by the same immanent end. They are both directed towards indetermination. Metabolism is the means by which organisms secure and maintain their individuality, but only so that they can expend stored energy in increasingly variable directions, through increasingly free acts. Reproduction is the means by which individuals are effaced in larger movements of change, so that life can develop in the direction of increasing structural complexity, each instantiation of which would secure and maintain itself as an individual through the metabolic canalisation of energy and its expenditure through free acts. Life in both its individual and evolutionary dimensions consists in this same pursuit of indetermination.

The rhythmic difference between metabolism and reproduction is therefore not really a difference of direction; it is a difference of time scale. The two tendencies only appear to be directionally antagonistic when one scale is taken as the frame for the other.

> The profound cause of this discordance [between the individual and the evolutionary] lies in an irremediable difference of rhythm. Life in general is mobility itself; particular manifestations of life accept this mobility reluctantly, and constantly lag behind. It is always going ahead; they want to mark time. Evolution in general would fain go in a straight line; each special evolution is a kind of circle. Like eddies of dust raised by the wind as it passes, the living turn upon themselves, borne up by the great blast of life. (CE 128)

Bergson adopts the evolutionary scale as the frame for the individual. The individual appears to lag behind the process of speciation, striving to retain its own form instead of giving way to evolution. The individual appears to be turned in on itself, forming an autopoietic circle, in contradistinction to the forward direction of the evolutionary movement. Bergson concludes that 'the act by which life goes forward to the creation of a new form, and the act by which this form is shaped, are two different and often antagonistic movements' (CE 129). Evolution via reproduction seems to aim at a constant change in form, while the individual seems to aim at persisting in the form that it has.

Yet Bergson adds that 'the first is continuous with the second' (CE 129). He likens the appearance of directional contrariety to the image of a man leaping to clear an obstacle but having to look at himself instead of the obstacle in order to clear it (CE 129). This man would appear to be concerned with himself instead of with the obstacle, but his self-directedness is only the means by which his aim to clear the obstacle is realised. There is, in principle, just one direction to the act. Only in fact does there appear to be two. The same could be said for the relationship between individuality and evolution. In principle, there is just one direction orienting the unfolding of biological processes on each scale. Only in fact, from the perspective of one of them, does there appear to be two. The evolutionary movement consists in a change of form ratcheted in the direction of increasing indetermination, but it unfolds through the medium of formally self-maintaining organisms. While organisms appear to be self-directed, it is their self-direction that furnishes the means through which evolution takes place.

The image of a vortex may be more familiar. Cuvier wrote that 'the living being is a whirlpool constantly turning in the same direction, in which matter is less essential than form' (qtd. in Gilbert 1982: 154).

Whirlpools or vortices maintain their forms through a steady flow of water. Picture a river whose course consists in a contiguous series of vortices. The downward flow of such a river would be much slower than the speed of its whirlpools. Take the circular diversion of water through which the vortices maintain themselves as an analogue for the metabolic canalisation of energy through which organisms persist. Consider the flowing river in which the vortices occur as an analogue to the lineage of some present population of organisms. According to this image, the metabolic maintenance of individuality is internal to the movements of reproduction and speciation that run through it. Just as the river's flow courses through the series of apparently stable vortices that populate it, in life the reproductive establishment of lineages is realised across the individual organisms of which they are composed. The two only appear as directional contraries because of a difference in time scale. The river moves more slowly than its whirlpools. The rate of speciation is slower than the metabolic processes through which the individuals of evolving species maintain themselves in form.

Metabolism and reproduction – the individual and the evolutionary – are directionally continuous movements oriented towards the advancement of indetermination. That said, it is no accident that Bergson consistently adopts the evolutionary as his frame of reference, a position from which individual organisms appear to 'counterfeit immobility so well that we treat each of them as a *thing* rather than as a *progress*, forgetting that the very permanence of their form is only the outline of a movement' (CE 128). Bergson's preference is for the progress, the movement. If 'life in general is mobility itself', it is on the evolutionary scale that life is most fully realised. The individual always comes to appear as a stable reality in fixed form with delimitable spatial boundaries via a temporary deceleration of the evolution- ary process. The individual seems for this reason more like the product and medium of evolution than its agent. This is a contention that strikes at the core of theories of evolution by natural selection.

Weismann redux

Bergson's apparent subordination of the individual to the evolutionary raises the question of the nature of the variations on which evolution, via selection, operates and depends. Bergson rejected Lamarck's idea that 'the experiences or behavior of the individual in the course of his career' could be material for selection, and agreed with the Neo-Darwinians that selection acts upon 'variations', or individual differences, inherent in the germ line or sex cells (CE 85). But he remained unsatisfied with explanations available for the cause of these variations as well as of their essential contingency, as accidents. 'I have spoken of selection as the paramount power', Darwin wrote, 'yet its action absolutely depends on what we in our ignorance call spontaneous or accidental variability' (1868: 236). Natural selection cannot operate without already-existent variations to select. It is derivative on variation's source. For Darwin, that source was chance, 'spontaneous or accidental' (cf. 1868: I, 9; II, 427–8; 1960: 177 and 1984: 284). By the 1920s, after the introduction of Mendelian inheritance into evolutionary theory, biologists began to invoke 'mutations in genes' as the 'source of new, stable variants on which selection can act' (Charlesworth and Charlesworth 2009: 761). That random mutation is the source of all variation is still mostly consensus today. One finds the following summation in a recent textbook.

> For a given population, there are three sources of variation: mutations, recombination, and immigration of genes. However, recombination by itself does not produce variation unless alleles are segregating already at different loci; otherwise there is nothing to recombine. Similarly, immigration cannot provide variation if the entire species is homozygous for the same allele. Ultimately, the source of all variation must be mutation. (Griffiths et al. 2000: 'Sources of Variation')

These contentions retain Darwin's commitment to variation's contingency. But some theorists have begun to express dissatisfaction with the idea, in large part because there has still never been a documented case of speciation by mutation, which means that its evolutionary efficacy as a source of variation remains undemonstrated (see Margulis and Sagan 2002: 72; Bergman 2003; Gerhart and Kirschner 1997; and Kirschner 2013). Bergson's issue with the purported accidentality of variation was not its empirical underdetermination, but its improbability. His case in point, as I mentioned, was the widespread convergent evolution of the camera-like eye (see Morris 2003: 151–7). He insisted that it was not adequately explained by sheer contingency and differential mortality alone.

Bergson saw in the Neo-Lamarckian doctrine of 'effort' a salutary alternative to Neo-Darwinism's directionless accidentality. According to the former, variation does not occur at random, but as the result of the way each organism's goal-directed activities, behaviours, and habits modify its physical structure (CE 76–7). This theory was based on Lamarck's postulation of 'internal nervous fluids' whose concentration relative to the use and disuse of parts in each individual was supposed to be conserved and transmitted across generations (Lamarck 1802: 9). By re-concentrating these fluids through the habit of stretching its neck to reach higher food, as in the famous example, a giraffe could make-heritable and transmit its strengthened and elongated neck to its offspring. A habit-based physical change becomes an evolutionarily salient variation. Bergson thinks that if the directional expenditure of effort can generate and conserve variations across generations, then the appearance of convergent traits can be better explained as an effect of the overlap in these directions, instead of as the repeated coincidence of random series of changes by chance.

The mistake of Neo-Lamarckism is its individuation of effort, or the restriction of variation's directional source to the individual scale (CE 86–7; cf. Kanamori 2010: 114–19). That there is directed effort beneath variation, Bergson affirms. But that this effort is of the individual, expended throughout the course of its own life, he denies (CE 85). On the latter point, Bergson sides with the Neo-Darwinians. Each theory corresponds to a certain aspect of the evolutionary process, correctly describing it only in part (CE 84, 87). 'It is up to philosophy', as one scholar writes, 'to disengage, at the intersection of the "trail of facts" traced by the sciences, the ideas that they suggest about the subject of life' (Caeymaex 2013: 53). From Darwinism, Bergson disengages the germinal understanding of variation from its accidentality; from Lamarckism, he disengages the directionality of effort from its individuation (see François 2008).

Since *Creative Evolution* was written in the wake of the Darwinian revolution but just before the Modern Synthesis of Darwinism with Mendelian genetics, its biological context is constituted by the way Weismann's work allowed Darwinism to shed its Lamarckian inheritance (see Ansell-Pearson 2005b: 61). Bergson turns to Weismann's theory of the germ-plasm to mediate between Darwin and Lamarck and secure his own explanation for the source of variation and its transmission as hereditary information (CE 26–7, 78–84).

Acquired characteristics are modifications made to an organism's physical structures through the habituated use or disuse of its parts. Organisms that make a habit of chasing a quick species of prey become faster predators with stronger legs. The idea that such an acquisition could be conserved and transmitted intergenerationally implies that the offspring of such organisms would be born having inherited stronger legs and a heightened capacity to run. Over a long enough series of generations, the effort to run faster might explain the gradual emergence of a new species, like the jaguar, from out of a slower ancestor in its family of Felidae, like the *Panthera blytheae*. The heritability of acquired characteristics is a straightforward instance of evolutionary finalism. Lamarck understood this theory – as did many others before him – to be a more or less self-evident implication of the progressive complexity evidenced across natural history (in the fossil record), together with the close conformation between the physical structure and habits of individuals (1802: 61; cf. Le Roy 1802: 227–8).

Lamarck's innovation did not consist either in the identification of the intergenerational conservation of acquired traits or in the articulation of a mechanism that would secure their heritability. Lamarck distilled from out of the ostensible fact of the transmissibility of acquired traits an agent of potentially unlimited change and a deep, directing principle central to the evolutionary drama as a whole (see Gayon 2006). Others – Darwin and Spencer, most notably – were more or less convinced, though they remained unsatisfied with the mechanisms for the transmission of traits available at the time (see Zirkle 1946). It was not until more viable mechanisms for heritability were discovered – with de Vries and Weismann – that debates surrounding the transmission of acquired traits took on a new polemical edge.

Darwin did not himself identify a viable mechanism for heredity. While he was able to account for the selection of variations – or difference – on the evolutionary scale, he was not able to account for stability – or repetition – on the reproductive scale. Canguilhem criticised Bergson for the same problem: lacking a material mechanism for heredity (the gene), Bergson was capable of accounting for difference, or the proliferation of forms, but only by leaving their generational stability or repetition over reproductive time unexplained (Canguilhem 2002: 339, 362). Canguilhem considered Bergson's epiphenomenal account of the stability of species to have been vitiated by the genetics of which Bergson was ignorant (see Feldman 2016: 157–61; cf. Talcott 2019: 113–24). Canguilhem took the development of gene theory to have uncovered a material basis beneath processes of transmission and mutation. The principle of the stability of forms is supposed to be secured through the structure of the gene, which is thought to account for the repetition of form by encoding it as information. If heredity is about difference and repetition, or variation and stability, then Bergson attained to only one half of the equation. But this is an unfair constraint. Bergson did not regard repetition, or stability, as a real feature of life. Stability of form is only the negative outline of a movement through it. As such, it does not require a principle of explanation. Heredity is not to be decomposed into difference and repetition, but into difference and continuity. Stability is better thought – recalling 'The Perception of Change' – not as the effect of repetition, but as a feature of relations between rhythms of continuity. Bergson likens the phenomenon to the illusion generated by parallel trains moving at the same speeds: the one appears to the other as if it were standing still (CM 119).

In any event, by the time of *Creative Evolution*, Bergson would have been at some remove from the initial debates surrounding the evolutionary efficacy of acquired characteristics. The primary theoretical issue consisted then in the theory's mechanistic viability on the basis of 'the supposed nature of germinal cells', the widespread acceptance of which was already supposed to have rendered Lamarckism inconceivable (CE 78–9). I mentioned earlier that the barrier established by Weismann between the germinal and somatic cells implies an inviolable causal independence between the modifications accomplished through habits – that is, the use and disuse of parts – and the germinal cells whose traits and their variations underwrite the development of the physical structures to be modified. On the germinal understanding, acquired characteristics are more or less strictly intransmissable. They occur in a secondary register, largely as effects of germinal properties and only rarely as their causes. Lamarck's finalism was supposed to have been mechanistically vitiated. Whether it makes sense to imagine physical structures varying with habit over time, the mechanism that would secure the conservation and transmission of that variation proscribes its very possibility.

Bergson accepts that 'the essential causes of variation are the differences inherent in the germ borne by the individual, and not the experiences or behavior of the individual in the course of his career' (CE 85). This is the Weismann barrier. It is not the only idea that Bergson recommends from germ theory. As much as he insists on the instantiation of variations in the germ, he tempers the mechanism of that claim with its opposite number. 'Where we fail to follow these biologists [i.e., the mechanists]', Bergson writes, 'is in regarding the differences inherent in the germ as purely accidental and individual' (CE 85). They are rather to be regarded as embodied in each individual germ while passing over and between them. Variations are better thought as 'the development of an impulsion' that is continuous across the germ line and trans-individual with respect to each particular germ. Individual differences should not be considered necessarily accidental. Their occasional convergent accumulation along divergent lines in the production of like organs should not be considered a feat of chance coupled with like environmental pressures. It could be that they are embodying in different individual germs one and the same line of development that is unified and continuous beneath them (CE 87).

Bergson evidently understands germinal *continuity* – alongside the individuality of germinal *difference* – to be a feature equally available to the reconstruction of Weismann's germ-theory (see Ansell-Pearson 2002: 94 and 1999: 151–2). He concedes that it might appear at first, at least under the 'extreme form' of the theory according to which 'the sexual elements of the generating organism pass on their properties directly to the sexual elements of the organism engendered', that Weismann is not entitled to the principle of continuity that he set out to establish (CE 26). This is because 'it is only in exceptional cases that there are any signs of sexual glands at the time of segmentation of the fertilized egg' (CE 26). There seems to be a gap in the transmission of germinal information between generating and generated organisms, an interval in which the latter develops the appropriate elements. The germ-plasm is not continuous, but has to be regenerated anew in each organism (CE 27). Bergson accepts this. He observes, however, that the sexual elements responsible for the continuity of germinal information 'are always formed out of those tissues of the embryo which have not undergone any particular functional differentiation, and whose cells are made of unmodified protoplasm' (CE 26–7). If it is true that germinal information is transmitted from the constituted sexual elements of the generating organism, then it must have been concentrated there initially, in the development of that organism. It seems as if there is, over the course of embryogenesis, an initial reception of germinal information in the egg, a subsequent distribution of that information over the differentiating tissues of the embryo (in order to direct their formation), and a re-concentration of 'something of itself on a certain special point, to wit, the cells, from which the ova or spermatozoa will develop' (CE 27). That re-concentration is facilitated by the fact that the ova or spermatozoa develop out of undifferentiated tissue. Bergson may be identifying the special quality of this form of tissue that later theorists will conceive in terms of its pluri- or equipotency (see Ruyer 2016: 57).

According to this reconstruction, Bergson is able to contend that while the germ-plasm is not itself continuous, 'there is at least a continuity of genetic energy, this energy being expended only at certain instants, for just enough time to give the requisite impulsion to the embryonic life, and being recouped as soon as possible in new sexual elements' (CE 27). The conceptual opposition between the individuality of germinal variation and the continuity of the germ line is resolved in terms of the distinction between the constitution of the germ-plasm internal to each individual organism and the genetic 'energy', or information, that impels the development of each germ-plasm while traversing it. 'Regarded from this point of view', Bergson concludes, *'life is like a current passing from germ to germ through the medium of a developed organism'* (CE 27). There is in Weismann both a principle of transmission that is individual as well as a continuity or current of energy, an impulsion, that 'indefinitely' pursues 'an invisible progress' and thereby organises the otherwise accidental individual variations of each germ along certain lines (CE 27; cf. Ansell-Pearson 2002: 94).

This 'progress' should not be conceived teleologically. 'Genetic energy' is characterised by ceaseless variation. It is constantly differing from itself in such a way that the new forms generated from out of it are unforeseeable in advance. It does not have those forms as its ends, for 'though the variation must reach a certain importance and a certain generality in order to give rise to a new species, it is being produced every moment, continuously and insensibly, in every living being' (CE 28). Variations do not arise accidentally in each individual as mutations, nor do they amass there as a result of goal-directed behaviour. Variations crystallise in individuals from out of a continuously modulating line that traverses them. Bergson conceives the mechanism of heritability, the germ-plasm, as the material instantiation and individuation of a continuum of energy that is, in itself, in a state of continuous modulation (CE 28). By understanding forms – individuals or species – to be derivative on a current that traverses them, Bergson recombines Neo-Lamarckism's directional variation with its Neo-Darwinian restriction to the germ line.

Orthogenesis

The idea that individual organisms are the medium through which a directional movement pursues a continuous development is one of the central tenets of the theory of orthogenesis. Orthogenesis – from the Greek *ortho*, meaning straight – is the position that evolution is governed by an intrinsic directionality (see Levit and Olsson 2006: 99, 130–2; and Bowler 1979: 40, 51). While Darwinian evolution consists in a stochastic process of variation sorted by the consolidating mechanism of selection, orthogenesis posits that variations are trended in definite directions, selection aside. Evolutionary outcomes are not wholly the consequences of the operation of a blind mechanism. This directionality could be a result of the imposition of a transcendent end as a guide for the evolutionary process, an inner drive impelling it forward from within, or the contouring effect of the pressure of external circumstance on the generation of individual differences. Common to each version of the position is the idea that variations are channelled in certain directions (whatever the reason), and that these trends define the shape of evolution in general.

Orthogenesis was first introduced as a term by Wilhelm Haacke in 1893, but it was Theodor Eimer's work on 'the impotence of natural selection' in accounting for such phenomena as the colour patterns of butterfly wings, five years later, that would prepare the term for popular uptake. Eimer distinguished between adaptive and non-adaptive orthogenetic trends, privileging the latter. That the consistent action of selection could generate the appearance of a directional development did not mark a break with Neo-Darwinian orthodoxy. This effect was eventually given the name of 'orthoselection' and incorporated into the Modern Synthesis. Eimer's emphasis on non-adaptive characters and their trends, on the other hand, represented a significant departure from the theory of evolution by natural selection.

Eimer began by observing, in line with Bergson, that even characters with adaptive utility could not always have had it (1898: 4, 22, 24–5, 31). Adaptive characters would either have had to develop through a process whose early stages were not yet useful, or else come to be out of the assemblage of previous variations, again not themselves originally adaptive for the later purpose. In many cases, adaptive value is a hypothetical postulate. Eimer's own field of animal colouration was populated with guesses about the utility of the colour schemes of shells and wings, usually understood as camouflage or for attracting mates. Such hypotheses are difficult if not impossible to confirm. They also leave unaddressed colours that are hidden from view, on the inside of a shell, for example. Eimer argued that such cases could be better explained by the postulation of inner forces whose effects could be stimulated and channelled by external pressures without necessarily consisting in an adaptive response to those pressures. One historian notes that 'in the end Eimer was so convinced of the prevalence of directed variation that he claimed there was no random variation at all' (Bowler 1979: 48). All of evolution consists in the contouring, via selection, of intrinsically directed lines of variation.

Eimer insisted nevertheless that orthogenesis remained a mechanistic theory of evolution. That evolution is a mechanical process should not require that we conceive its unfolding stochastically. The mechanism for orthogenesis was supposed to consist in the action of external constraint on internal powers. These powers are what manifest themselves in, and account for, the developmental patterns – what Eimer called the 'laws of growth' – of colouration, leaf-formation, and other morphological processes. These laws of growth are, as Bergson says, 'merely physical and chemical' (CE 74). They are part of the material composition of organic bodies. Each species is defined by the set of internally directed possibilities for variation available to it because of its constitution. As external pressures – environmental circumstances, climate, available nutrition, predators, and so on – gradually modify the physical constitution of a species, its 'laws of growth' are constrained and modified as a result. These laws are not objects of selection, since selection can act only upon characters presently manifest. The developmental laws that govern the manifestation of those characters are canalised and constrained as a secondary effect. Eimer attempted to subordinate the efficacy of selection to the inner directionality of variation without relinquishing the mechanicism of the theory as a result.

It is from Eimer that Bergson inherits his own understanding of orthogenesis (CE 72–6, 86–7; cf. François 2010: 63–77). When Bergson rejects the theory for its mechanicism, it is Eimer's physicochemical determination of variational tendencies that he has in mind (CE 72–4). Today orthogenesis is usually scorned among biologists and philosophers of biology alike (see Larson 2004: 127). Ernst Mayr's famous disparagement of the term in 1948 for implying 'some supernatural force' remains a popular response. George Simpson described it as involving a 'mysterious inner force', and most invocations of orthogenesis have linked it to vitalism in one way or another ever since (see Ruse 1996: 447). If directionality were the result of a vital force governing the evolutionary process, then Bergson would, of course, have to abandon the theory of orthogenesis for that reason as well. But this is not how Bergson understands Eimer. Variation is intrinsically directed. It is directional at its source. It is not directed by a principle transcendent to it – of which any vital force would be an example. Bergson considers the idea of immanent directionality to be orthogenesis' kernel of truth (CE 86). He seeks to disengage it from its basis in 'the mechanical composition of the external with the internal forces' (CE 76).

Orthogenesis is supposed to be clearest in cases of convergent evolution. Bergson observed that the same organs are converged upon not only evolutionarily, but developmentally or ontogenetically as well. In Chapter 1, I mentioned that the evolutionary histories of sapiens and cephalopods eventually arrive at the same form and function of the same organ, the camera-like eye, 500 million years after diverging from the last ancestor they had in common (CE 74–5; cf. Ogura et al. 2004). This is phylogenetic convergence. The eye is not only converged upon in the evolutionary event when two distinct species arrive at the same complex result. The vertebrate eye is formed out of brain tissue, while the cephalopod eye is formed out of the epidermis (see Serb and Eernisse 2008). Thus the eye is converged upon developmentally as well. 'Every moment, right before our eyes, nature arrives at identical results, in sometimes neighboring species, by entirely different embryogenic processes' (CE 74–5). This is ontogenetic convergence. If the 'production of the same effect by two different accumulations of an enormous number of small causes is contrary to the principles of mechanistic philosophy', then more than phylogenetic convergence, it is ontogenetic convergence that presents the strongest challenge for mechanist biology (CE 76). For 'here we have, indeed, the same effect', as Bergson says, 'obtained by different combinations of causes' (CE 76).

Eimer was right to base a theory of orthogenesis on the directional character of variation, to which cases of convergence seem to testify, but wrong to suppose that directionality could be explained via the concatenation of mechanical causes alone (CE 86). Bergson affirms that 'variations of different characters continue from generation to generation in definite directions', but denies that 'physical and chemical causes are enough to secure the result [of the definite directedness of evolution]' (CE 86). Orthogenesis should not be conceived mechanistically. It should not be conceived finalistically either. Opposite the physicochemical determination of directionality is the postulation of ends prior to the evolutionary processes that they would work to direct. Bergson rejects this possibility as well. Whatever accounts for orthogenesis cannot be taken to reside outside of and beyond the evolutionary process itself, either as transcendent to or as precedent of it.

According to philosopher of biology Pete Bowler, Bergson's position on orthogenesis should be understood in the finalistic sense that evolution is directed by a force that resides outside it (1992: 57). The evolutionary process takes the form of a linear trajectory tending towards some limit, after which point it is divided, diverted, and linearised over again (1992: 116–17). Though there is no predetermined goal to orient and explain the direction of evolution, there is nonetheless something operative within evolution that is transcendent to it. Evolution owes its orthogenicity to this transcendent x. That is Bowler's reading. I think it is half wrong. What it gets right is the idea that directionality is an effect of a tendency in the process of realising itself in successive phases towards a limit. What it gets wrong is the idea that this tendency is transcendent over the evolutionary process, as well as the individuals through which it is unfolded. I return to this misconception in the next chapter. I show there that the appearance of the externality of the *élan vital* to the evolutionary history within which it is operative is an effect of the distinction between the process and product perspectives on development that I discussed last chapter. It is only when evolution is conceived from the perspective of product that the principle of its development appears to be something other than the process itself.

Bergson posits within the evolutionary process a qualitative multiplicity of tendencies. They are what account for the directionality of evolution. Tendencies tend towards limits. Their limits are immanent to the tendencies themselves, being nothing other than how far they can go. If it does make sense to speak of a Bergsonian theory of orthogenesis, then it should be on the basis of the immanent directionality

of tendency. The limits that define the tendencies at work in orthogenetic lines of development do not precede those lines, nor do they lie
above and outside of them. They are immanent to – coextensive and
coterminous with – the evolutionary trajectories whose directedness
they explain.

Vitalism in dispute

It may be objected that Bergson's variation on orthogenesis is a classic
example of vitalism and should be rejected on that basis. Vitalism is a
controversy in a word.[2] Before it was minted as a noun, the term was
coined in adjectival form by Charles-Louis Dumas in 1800 to refer to
the Montpellier medical school in France.[3] This school represented a
stronghold against Descartes' iatromechanical model of the living body,
which regarded organisms as complex machines and sought to describe all
bodily processes in mechanical terms. Against the mechanistic view, the
Montpellier doctors maintained a distinction between organised bodies
and inert matter, arguing that life could not be adequately explained by
mechanical principles alone.

As Bergson recognised in his early lectures on the subject (1990: 337–
50), the school was also initially opposed to animism, which the doctors thought went too far in the other direction by regarding all organic
processes as a function of mind or soul. The preeminent animist of the
eighteenth century was Georg Ernst Stahl. Stahl conceived the body and
its organs as the instruments of the soul. François Boissier de Sauvages
later introduced Stahl's idea of the soul as an explanatory principle into
the Montpellier school, mediating between an animistic understanding
of organic self-maintenance and the mechanical explanations popular in
medicine at the time. By the second half of the eighteenth century, two
broad notions had come to characterise the Montpellier approach. One
was the idea of organisation, or the relationship between organic parts
and the irreducible whole that their interactions produce. The second
was the idea that living bodies are characterised by vital properties such as
irritability and sensitivity, which cannot be explained in terms of the laws,
methods, and principles of lifeless matter. Theophile de Bordeau argued
that organisation was autonomous from and irreducible to the mechanics
of inert matter. Eventually, that view attached itself to Paul-Joseph
Barthez, whose work Dumas had in mind when he famously referred to
the Montpellier doctors as vitalists.

Vitalism, according to the retrospective caricature, is supposed to
consist in the idea that life is not only different from matter, but cannot
be understood on the basis of matter, being irreducible to it. According

to the influence of animism, life's irreducibility is understood in terms of something superadded to the material world, as matter on its own is incapable of self-organising, self-maintaining, or evolving. Life must be matter + x. Different vitalists have tended to determine that x through a series of non-mechanical, occasionally immaterial, uniquely vital principles, forces, substances, or properties, irreducibly vital facts that would account for the special qualities of organisation.

These are generally considered to be discredited ideas. Vitalism is supposed to have been refuted twice (see Wolfe and Wong 2014: 66–7). In 1828, Friedrich Wöhler synthesised urea, demonstrating that organic substances can be produced out of inorganic compounds and undermining the idea that organisation could not be reduced to or explained by the chemistry of inanimate matter (see McKie 1944 and Ramberg 2000). Roughly a century later, the Vienna Circle argued that physics had disclosed a causally closed material universe, invalidating any claim to an immaterial force that could causally impact material bodies without being describable in material terms itself (see Frank 1998). Vitalism, if it continues to insist on the inadequacy of chemistry to organised matter or advocate for the material efficacy of an immaterial force, seems to lack any scientific credibility.

This may be the dominant view among scientists today. Ernst Mayr's statement on vitalism is a popular touchstone for biologists. Vitalism 'virtually leaves the realm of science by falling back on an unknown and presumably unknowable factor' (1982: 52). Carl Hempel's criticism is a popular touchstone for philosophers of science. The issue is not that vitalism posits unobservable entities, but that these postulates 'render all statements about entelechies inaccessible to empirical test and thus devoid of empirical meaning' (1965: 257). Philosopher of biology David Hull concludes, with reference to vitalism, that 'both scientists and philosophers take ontological reductionism for granted. Vitalism is dead. Organisms are "nothing but" atoms, and that is that' (1981: 282).

Yet recent scholarship has made considerable headway in complicating the picture by attending to earlier and subtler forms of materialism and by distinguishing between different types of vitalism and drawing out the heuristic or scientific utility of some of them. The history of materialism is not the history of a progressively exhaustive elimination of vital facts, forces, properties, or principles from a law-bound material world, fully determinable in reductively mechanistic terms alone. The history of vitalism is not the history of an increasingly implausible series of speculative postulates made with dogmatic certainty about the inadequacy of the natural sciences to the transcendent properties of life. Vitalists usually identified as vitalists as a reaction to what they perceived as inadequacies in the reductionist, materialist, and mechanist theories of their time. Many vitalists were accomplished experimentalists in their own right. They subjected their views to empirical testing and produced results that helped advance research in their fields. In a number of cases, the shortcomings that they identified in the prevailing forms of mechanism were verified empirically and served as incentives to produce better and more complex explanations. It is the constructive hypotheses regarding the nature and existence of irreducibly vital phenomena that the sciences have consistently supplanted.

a. Entelechy

By the time of *Creative Evolution*, the vitalism controversy had mostly settled around the experimental embryology of Hans Driesch. Driesch is the primary target of Mikhail Bakhtin's infamous 1926 essay 'Contemporary Vitalism', but Bakhtin includes Bergson as part of the neo-vitalist school, what Bakhtin calls 'critical vitalism' (1992: 80–2, 96; cf. Wyk 2012 and Gruner 2017). Bergson mentioned Driesch in a footnote to *Creative Evolution*'s brief critical discussion of vitalism. Driesch referred to Bergson repeatedly and wrote a positive review of Bergson's book. The two have been considered together under the banner of 'neo-vitalism' ever since (see Sumner 1916; Freyhofer 1982; Vucinich 1988: 169; Midgley 2011; Garrett 2013; and Bianco 2019; cf. Posteraro 2021a). This is a mistake. To demonstrate it, I present a brief overview of Driesch's vitalism. Then I show that Bergson had already provided a sophisticated criticism of it.

Driesch's first major theoretical work in English, *The Science and Philosophy of Organism*, appeared in 1908, drawing heavily from earlier works published in German. It offered a defence of vitalism on the novel basis of the experimental facts of regulation and regeneration. Driesch's idea was that only if undisturbed development was possible could everything about organisms be mechanistically explained (1929: 103). What he set out to prove was that development could be interrupted without the individuality of the developing organism being compromised as a result. Cases of regulation and regeneration evinced the point. Driesch argued that they could not be understood mechanistically, and that they testified to the non-mechanical action of an immaterial force.

By compressing early sea urchin embryos between glass plates, Driesch was able to reconfigure the divisions in their eggs, reshuffling their nuclei so that some that would normally have produced dorsal structures were found in ventral cells instead. According to mechanistic (or preformationist) principles, the embryos should have developed in a disordered and unviable fashion. Yet Driesch famously obtained normal larvae from them, which meant for him that the early embryo was composed of pluripotent cells, and that the developmental processes through which they gave rise to differentiated organs must be self-regulating. Driesch linked the phenomenon of regulation to the already well-established facts of regeneration such as it occurred in his own experiments on salamanders, which are capable of regenerating the lenses of their eyes after they are removed.

Driesch argued that regulation and regeneration indicate the existence of an individualising agency at work in the organism, distinguishing it from the mere mechanical assemblage of parts and securing its autonomy as an organised whole over and above changes in its constituent elements. Driesch's word for this agency was 'entelechy', an Aristotelian term with its roots in the Greek *enteles* (complete), *telos* (end), and *echein* (to have). Leibniz popularised the word with the definition of 'something analogous to soul, whose nature consists in a certain eternal law of the same series of changes, a series which it traverses unhindered' (1989: 173). Driesch's entelechy is an immaterial force, acting to bring about the unified development of an organic individual from out of initially pluripotent cells (see Chen 2018 and Bognon et al. 2018). As a result of entelechy, 'a sum (of possibilities of happening) is transformed into a unity (of real results of happening) without any spatial or material preformation of this unity' (Driesch 1929: 215).

Entelechy was what guided initially pluripotent cells to the specific structures in which their development culminated. Cellular pluripotency explained the fact that cells isolated at the two-cell stage of development in sea urchin eggs did not produce two half-embryos but two fully formed organisms. Entelechy explained the way those cells were guided towards their final forms, since their pluripotency seemed like evidence of the idea that no physical or chemical structures existed in order to determine development in advance. Driesch thought that cases of regeneration also supported the existence of entelechy, since they demonstrate the way the individuality of the organism could be safeguarded against changes to its composition. Not only can living things self-regulate developmentally, but they can do so compositionally as well (1929: 153–4). The special force that brings about the individual whole from out of a pluripotent cellular field secures the integrity of that whole once it is constituted. Regulation is regeneration for the adult organism. Driesch supposed both to be impossibilities for mechanical systems. He concluded that 'embryological becoming is "vitalistic" . . . it is impossible to comprehend it by the laws of physics and chemistry' (1914: 226). This is vitalism in its classical formulation (see Van der Veldt 1943, Klaus 1997, and Sapp 2003). Physics and chemistry are inadequate to the explanation of biological phenomena. Biology requires the addition of a supplemental principle.

As Driesch demonstrated, vitalisms consist in both critical and constructive elements. Their criticisms target the scientific understanding of matter, usually mechanistically conceived, and argue that it is insufficient to the explanation of what is distinct about biological phenomena. Their constructive arguments advance varying positions regarding the new and irreducible principle, property, or force that has to be introduced to capture the specificity of life. Vitalism's critics purport to attack both, but it is only really a succession of variants of vitalism's constructive aspect that have been consistently discredited. Like many others, Bergson considers the critical moment worth taking seriously (CE 42, n. 1). Life – organisation and evolution – is irreducible to physicochemical explanation. In this respect Bergson is no doubt a vitalist in the critical sense. 'The vital principle may indeed not explain much, but it is at least a sort of label affixed to our ignorance, so as to remind us of this occasionally, while mechanism invites us to ignore that ignorance' (CE 42). Vitalism may serve a heuristic function, regulating against mechanism's tendency towards the reduction of biological complexity. That should not go unnoticed. But the more pressing question is whether Bergson advances his own constructive position, of the sort that would put him in line with Driesch.

b. Individuality

Bergson does of course advance a positive theory of his own, but it is not one that puts him in line with vitalists of the traditional variety. Bergson was in fact a critic of them. Alongside Driesch's 'entelechy', Bergson also refers to 'dominants' (CE 42, n. 1) He comments at some length upon the deficiencies of Claude Bernard's ideas concerning the existence of 'vital force' as well (CM 241–2). Vitalist postulates share on Bergson's account an important deficiency with mechanistic models. Both are human contributions and do not exist in nature independently. According to this criticism, Driesch's entelechy is an intellectual abstraction born of the projection of the manufacture model of organisation onto the biological world. It understands organisms as if they were built artifacts and attempts to explain their composition on that basis.

> When we think of the infinity of infinitesimal elements and of infinitesimal causes that concur in the genesis of a living being . . . the first impulse of the mind is to consider this army of little workers as watched over by a skilled foreman, the 'vital principle', which is ever repairing faults, correcting effects of neglect or absentmindedness, putting things back in place. (CE 225)

This first impulse is natural to the intellect, a product of our adaptation to acting on matter. It consists in treating the organism as if it were an object, its organisation as if it were designed, and concluding that there must be a principle to account for that design, just as artifacts are constructed and repaired by external agents. The mistake is in thinking that organisms are complex in the same way that made things are complex. The appearance of that complexity is only 'the work of the understanding' (CE 250). It is not a fact, but a projection, and so does not require a superadded principle that would act as a designer in order to explain it.

Anthropocentric artifactualism is the first problem with any vitalism that accounts for organisation through the postulation of an organising principle. The second problem is the supposition that determinate individuality is a biological reality (CE 43). Organisms are composed of a multiplicity of elements that are each organised in their own right. If the vital principle is a requisite condition of individuality, then each organ of each organism and each cell of each organ should warrant vital principles of their own, to say nothing of the symbiotic relations that contaminate the boundary between the organism and the closely allied forms of life that live both inside and outside of it. The postulate first made to account for individuality eventually works to undermine it. If organisms are not the kinds of coherently self-sufficient individuals to which one could reasonably ascribe individualising principles, then they do not instance the kind of autonomy required to accommodate such principles either. 'The individual is not sufficiently independent, not sufficiently cut off from other things, for us to allow it a "vital principle" of its own' (CE 42).

Organisms are incomplete metaphysical individuals from the reproductive perspective as well, understood as moments in a history. The adult organism 'is only the development of an ovum', and the fertilised egg, inasmuch as it is formed by 'part of the body of its mother and of a spermatozoon belonging to the body of its father', is 'a connecting link between the two progenitors' (CE 43). Those progenitors are continuous with their progenitors in turn. Thus 'each individual may be said to remain united with the totality of living beings by invisible bonds', that is, historical ones (CE 43). Vitalism, if it is supposed to take organised individuality as its datum, would seem to have to collapse back onto the entire history of evolution, encompassing 'the whole of life in a single indivisible embrace' (CE 43).

Bergson intends this claim to work both critically and constructively at the same time. If by vitalism we understand the postulation of a principle or force internal to the purposive organisation of living things in order to explain their irreducible distinction from inorganic matter, then

Bergson is not a vitalist. There is no autonomous individuality in the organic domain. Yet Bergson does not conclude by rejecting the idea of irreducible vitality. He affirms instead that if it is to be attributed to the organic domain, then it has to be predicated of the evolution of life as a whole, as a single event in the making.

Evolution, as a single event, is supposed to be unified by the unfolding of tendency. 'Life is tendency' (CE 99). Bergson's infamous name for this tendency is the *élan vital*. It is the *élan vital* that attracted the vitalism controversy to him. According to its critics, the *élan vital* is like entelechy, a vital force or principle, immaterial, maybe even spiritual, and unambiguous evidence of a scientifically outdated philosophy. But it is a mistake to understand the *élan vital* as an organising life-force acting to explain and secure the organisation distinctive of life (TS 112, 115–16, 249). It is not a principle superadded to the material complexity of living things in order to account for their specific unity. It does not correspond to biological individuals, generating and safeguarding their autonomy in contradistinction to the mechanistic behaviour of inert matter. The problem with vitalism is not that it insists on a difference between life and matter, but that it misconceives each term of the binary and then incorrectly individuates the difference-maker between the two. Bergson's difference-maker is not a life-force but a tendency, and it is external to the ostensible individuality of any given biological form. Driesch's entelechy safeguards biological individuality while Bergson's idea of life is intended to explode it.

It may be the case that Bergson's idea of tendency involves aspects of vitalism. In any event, I think it is better understood as a form of finalism instead, an immanent and inverted form of finality in particular. Partially individuated living things are unified from behind, via the directional unfolding of the evolutionary process. The *élan vital*, as I show later, is Bergson's image for this unifying process. It is the finality of the process, as Bergson conceives it, that makes his theory distinctive. If the theory is to be correctly assessed, it should be on the basis of the idea of finality, not according to the traditional accounts of vital forces that Bergson distinguished his own theory from.

True finalism

Finalism, as it has been traditionally conceived, undercuts the self-sufficiency and temporal reality of evolution by positing ends prior to and transcendent over the evolutionary process in order to explain the directions it takes. In Chapter 1, I mentioned Bergson's view that mechanism cannot be rejected without some form of finalism taking its place

(CE 40). In this case, it is an inverted form of finalism that takes the place of Eimer's mechanism in Bergson's reconception of orthogenesis. Bergson calls it a 'true finalism' (CE 52). I think it is the best way to understand his account of vital tendency and the most appropriate background for a discussion of the *élan vital*.

Following Bergson's critique of internal finality, the renewed form of finalism has to be understood as a variant of external finality. It is made up of three components: (1) a reconception of the 'whole' to which externalism attributes purposiveness; (2) an inversion of the location of that purposiveness within the whole; and (3) a reformulation of what it is that accounts for the existence of that purposiveness in the first place.

As I showed above, Bergson's argument against internalism turns on his denial of the complete reality of organic individuation. Here is the outline of the argument as I reconstructed it in Chapter 1.

1. Inner purposiveness requires determinate individuality
2. There is no such individuality in the organic domain
3. Thus, there is no way to locate finality *internal* to any biological individual
4. Thus, if there is to be finality, it must be *external* to all biological individuals

Bergson concludes that 'finality is external or it is nothing at all' (CE 41). Internality requires that divisions be cut into the organic domain. Since they are always incompletely determinate, the locus of finality will always reside external to them, on the outer side of any individual so considered. To say that finality is a necessarily external attribution is to say that it has to be attributed to all of life indivisibly (CE 43). Finalism should qualify the whole from out of which internalism attempts to dissociate individual parts.

The problem with externalism is its recourse to the metaphysics of possibility. Externalism posits a set of pre-existent possibilities and conceives evolution as the progressive actualisation of them. Finality, on that account, is external to any individual form because it is located in a modal register alternative to the domain of actual individuals. But it is a modal register that effaces the temporal reality of evolution. The form of finalism that Bergson endorses qualifies life as a whole, but locates the finality of the whole in a modal register other than that of the possible.

a. Externality

Bergson's idea of an evolutionary 'whole' does not refer to the sum total of all currently present individuals (actualism), or to the concept of a pre-existent plan on the basis of which they would be actualised over time (possibilism).

> If there is finality in the world of life, it includes the whole of life in a single indivisible embrace. This life common to all the living undoubtedly presents many gaps and incoherences, and again it is not so mathematically *one* that it cannot allow each being to become individualized to a certain degree. But it forms a single whole, none the less; and we have to choose between the out-and-out negation of finality and the hypothesis which co-ordinates not only the parts of an organism with the organism itself, but also each living being with the collective whole of all others. (CE 43)

The idea of the whole is the idea of commonality across life. Is there anything that unites every living being with every other at every level of organisation? Bergson's explicit answer comes late in *Creative Evolution*. That which 'links individuals with individuals, species with species, and makes of the whole series of the living one single immense wave flowing over matter' is, he says, the 'unity' of 'the *élan*' 'passing through generations' (CE 250 tm). He means the unity of tendency. This unity is a 'movement', 'a simple process' (CE 250–1). Partially individuated organisms are the moments of a movement or phases of a process that traverses and realises itself across them. That movement, that process, is what all life has in common (CE 128). The whole of life is the movement of evolution considered as a single unfinished event, driven by a unified impulse (see Montebello 2012a and Miquel 2008).

Bergson seems to consider the unity of life ultimately impossible for the human intellect to represent completely (CE 49–50). We can draw a provisional set of instructions for how to imagine it. We should begin by placing 'ourselves in one of the points where evolution comes to a head', the body of *Homo sapiens*, say (CE 49). We should consider this form not as a present and complete state of affairs, but as the most recent phase of an evolutionary history that 'had to abandon by the way many elements incompatible with this particular mode of organization and consign them . . . to other lines of development' (CE 49). Then we should imagine each 'diverse and divergent element', the other forms striated along the other lines of development, 'as so many extracts which are, or at least which were, in their humblest form, mutually complementary' (CE 50). We should resituate the form of the human body within a larger nexus containing all the other forms arrived at by all the other evolutionary lineages. Finally, we should consider this immense mass of elements not as a result, a collection of evolved forms, but through 'the act by which the result is obtained', the evolutionary process of the formation of each form over time and from out of each other, reaching all the way back to the common ancestors of all life (CE 50). This is the closest we can come to conceiving of life as a whole. But even here, we cannot help but restrict to its present stage a process that is as yet unfinished, still unfolding (CE 50).

Consider another image, a finished sculpture – Michelangelo's David. Picture the process through which a massive marble block was quarried from an Italian mountainside in the Apuan Alps, cut down, chiselled, chipped, and sculpted into finished form. David is the terminal phase of a process that culminated in him by progressively selecting away everything that was incompatible with his figure. Try to imagine the other artifacts that were sculpted from out of the same marble slab, cut from the same quarry, harvested from the same mountain, all the busts and tiles and altars and facades. Think back even further to the formation of the marble itself hundreds of millions of years ago. Imagine the geological eons over which generations of micro- and macro-organisms were fossilised, layered and compressed at the ocean's floor, congealing and metamorphosing into an interlocking mosaic of carbonite crystals. Imagine the tectonic processes through which the ancient floor rose up from out of the ocean and into mountainous arcs over southern Europe. Imagine being there, present, able to witness the rise of marble from the sea. Already the event in front of you was carrying within it a deep history of development. You could not imagine the transformations it was still to undergo. You might be able to guess that eventually it would be harvested for purposes of craft, but you could never foresee the ramifying lines of alteration spanning from the white stone through the history of the plastic arts and into the production of the statue of David. Each moment in that history, from the biogenic formation of minerals to the harvesting of marble to the sculpting of a statue, is to each living being what the history in its entirety is to the whole of life. At every one of its moments it is a history unfinished and impossible finally to foresee.

Life as a whole is external to any one of the ostensibly individuated biological forms that populate it because individual forms are only ever artificially stable perspectives on what is an event in the process of unfolding itself (CE 128). To invoke another image, one of Bergson's favourites, consider the flight of Zeno's arrow. The localisation of the arrow at any of the spatial locations through which it passes is an artificial operation because those locations are spatialisations of a qualitatively whole movement, and have no reality outside of it. The arrow's trajectory is external to any one of its possible locations in space as those locations represent possible stopping points and are therefore derivative on the movement itself. The movement is not only external to its possible stopping points; it is also immanent to and coextensive with them. The same should be said of life. Life is the qualitative whole of the event of evolution. Determinate organic forms are only its possible stopping points, and evolution is therefore external to any possible set of them. At the same time, and as a result, evolution is also immanent to all determinate forms, for they are nothing outside of the movement through which they are formed (CE 43–4).

The modal status of life's unity is virtual. The whole is never given in its entirety at any actual (or possible) moment. The whole is the condition for the generation of each of its moments. Because it is always unfinished, incompletely realised in any possible present configuration, it is a whole that is essentially open. Over the course of evolution, the unified whole of life is progressively dissociated and enumerated across a plurality of actual determinations. It is instantiated in each of its parts, or partially individual organic forms, without being reducible to the sum total of them. Every living being represents a relatively stable phase in the happening of the one event of life, evolution.

The idea that every organism instantiates the virtual whole means that every organism is ultimately indissociable from the entire history of evolution. Each organism retains the virtual whole of the event as the evolutionary past within it. It is one part of the present of that history. According to Bergson's virtual mereology, the whole also implicates each of the parts that instantiate it. It does not contain them in the sense of a quantitative multiplicity of prefigured parts, the way prespecified possibilities exist in the register of the possible before becoming actualised. The whole implicates its parts in the state of a qualitative multiplicity of tendencies, the way a slab of marble, because of its composition, tends towards the realisation of a range of potential shapes and structures without resembling them in advance.

Variations on finalism are determined and differentiated from each other in two ways. Chapter 1 examined the first under the category of object-individuation, the specification of *what* it is that a given theory is predicating purposiveness of. The second consists in the question of *where* purposiveness is located within a given object, process, or system. When the object at hand is a built artifact, external finalism locates its purpose, diachronically, at the end of the process through which the artifact was assembled. The end functions as an ideal aim that directs the integration of initially disparate materials over the course of construction. The process is complete when the end is realised in the terminal unity of the built object. Bergson accepts externalism's individuation of the object – the whole of life – but relocates the purposiveness.

b. Commonality

External finality is teleological. It derives purposiveness from the terminal end, or telos, of a given process. Bergson's variation on finalism inverts the teleology. The location and realisation of ends are indicated by the presence of a harmonious or unified relation among an otherwise heterogeneous set of parts, as in the case of the artifact. Harmony, or the unity of a whole, is a function of the realisation of purpose. Disharmony, or the disintegrated plurality of parts, is typically an indication that some relevant purpose has either yet to be approximated or is being moved away from. 'If life realizes a plan', as external finalism supposes, 'it ought to manifest a greater harmony the further it advances, just as the house shows better and better the idea of the architect as stone is set upon stone' (CE 103). The purpose of the house resides in its end, which is given at the outset in the mode of a possibility to be realised, acting as a guide for its own realisation. Finality, understood as the realisation of a plan, supposes that the initially 'many' become an eventual 'one' (CE 92).

This is a bad model for development and evolution alike. It gets them backwards. 'Life does not proceed by the association and addition of elements', as does artifactual construction, 'but by dissociation and division', that is, by differentiation (CE 89). Development proceeds from the 'one', a unified equipotential zone, through progressive differentiation towards the 'many', the structural complexity of the adult organism. According to Bergson's critique of the bounded individuality of the organism, the plurality of parts into which development culminates should not be thought of in terms of the sort of unified and harmonious system to which purposiveness could be ascribed. If there is harmony or unity anywhere in this process, it is to be located in the originating condition of development. Whereas the artifactual model takes for granted the integration of a set of distinct elements in view of a common end, development operates according to the differentiation of an initially unified multiplicity *away* from a common origin.

Bergson makes the same argument for evolution (CE 40, 103–4, 127). It appears that the evolutionary process, far from testifying to the global harmony attendant upon the realisation of a plan, actually manifests an increase in *disorder* over time. There are two main reasons why. First, if life pursues an evolutionary progress, it does so across divergent lines of speciation. It is in the essence of tendency to extend itself as far as it can; once it reaches its limit, 'it divides instead of growing' (CE 99). Tendencies divide in order to continue developing, along different lines, past a threshold that marks the fullest possibility of one of them. The essence of an evolutionary tendency is 'to develop in the form of a sheaf, creating, by its very growth, divergent directions among which its impetus is divided' (CE 99). Second, nature is not coherent (CE 104). Taken together, the divergent directions through which species evolve are not complementary, but discordant. Their unity lies behind them, in the condition away from which they have developed. Ahead of them lies widening separation. 'Life, in proportion to its progress', as Bergson writes, 'is scattered' across 'mutually incompatible and antagonistic' forms (CE 103). Unsuccessful species represent the arrest of any semblance of evolutionary progress, while selection pressures oblige successful species to specialise themselves more closely to their particular conditions of existence, selecting and shaping their niches just as they are selected and shaped by them, deviating further from each other in the process (CE 104). Bergson concludes that 'the discord between species will go on increasing' as evolution unfolds (CE 103).

It does not follow that there is no unity or harmony to be found in the evolutionary domain. Bergson draws a different conclusion. 'Harmony is not in front' of the evolutionary process, indexed to a pre-existent end, 'but behind' it (CE 103). 'It does not exist in fact; it exists rather in principle' (CE 51). According to the first law of tendency, the law of dichotomy, for any set of divergent and conflicting tendencies, there was prior to their divergence a primitive tendency that contained them both (TS 296). The converse of this law implies that every pair of conflicting tendencies has its genetic condition in the bifurcation, or 'splitting up', of an antecedent tendency (TS 296). This is something like a metaphysical formulation of Darwin's tree of life, as I mentioned in Chapter 3. Speciation operates through ramification. One line of descent divides into two, each of which eventually branches again. Every lineage shares a common ancestor with every other, reaching all the way back to the emergence of the first living being – supposing, of course, that life originated on earth just once. The currently diverse array of extant life forms, many of which are incompatible with and antagonistic to each other, is ultimately unified in the history of evolution as a whole. Unity is a point away from which life diverges, not an end into which it eventuates.

Bergson's alternative to Darwin's tree is an artillery shell. Evolution does not look like a 'solid ball shot from a cannon', following a single linear trajectory, but 'rather like a shell, which suddenly bursts into fragments, which fragments, being themselves shells, burst in their turn into fragments destined to burst again, and so on for a time incommensurably long' (CE 98). Bergson favours this image for what it suggests about the ramifying evolution of life (see Jankélévitch 2015: 122; cf. Bernet 2012). 'When a shell bursts', he writes, 'the particular way it breaks is explained both by the explosive force of the powder it contains and by the resistance of the metal' (CE 98). There are two series of causes at work: one internal and expansive, the powder's own explosive force, and the other external and restricting, the constraint of the metal shell. The powder is what accounts for the shell's outward movement. The metal is what accounts for its fragmentation. The image of the shell correlates differentiation with external constraint and commonality with inner impulsion. The shell fragments because of the material resistance encountered by the powder's explosive force. What each fragment shares in common is what impels it through further fragmentations in turn, that is, the force of the original explosion.

Life carries its own explosive force within itself due to 'an unstable balance of tendencies', a multiplicity that serves as evolution's 'motor principle' (CE 98 and 101). This force is constrained and contoured by the resistant force of adaptation exerted through environmental pressure and manifested in selection (CE 98). The interaction between the two forces is what explains ongoing differentiation. Life is propagated outward on account of its own inner tendencies, and that outward movement is differentiated across distinct and divergent lines of development on account of the external pressures encountered by the originary impetus of a tendency tending to its limit (CE 102). It divides in response to the external obstacles that would otherwise arrest its forward movement (CE 99). What each divergent direction shares in common is what continues to impel it onward until it encounters resistances that force it to divide and differentiate again, that is, the prolongation of the original impulsion (CE 100–1).

The identification of commonality with impulsion is fundamental to Bergson's reformulated finalism. I said that harmony is manifested as a function of purposiveness. Teleological finalism locates harmony in the completion of a process of change. It locates purposiveness in the end towards which the process is oriented and into which it culminates. The end operates as an attractant, pre-existing its realisation and exerting its influence on the process from ahead. Bergson inverts the first of these precepts. There is unity or harmony in the domain of life – whether developmentally or evolutionarily determined – but it is located at the origin of a process of change, as change means differentiation and differentiation means directional divergence. It is never manifest in fact, but only in principle, as the state away from which evolution is always in the process of developing (CE 51). If pre-existent ends unify *initially* disparate elements by *attracting* them from *ahead*, then common origins unify *eventually* disparate elements by *impelling* them from *behind* (CE 103). That is the inversion.

The result is an immanent finalism. Life 'takes directions', but 'without arriving at ends' (CE 102, 16). Life is purposive because it is directional. Its shape is not entirely the result of a series of accidents pressed into form via the mechanical force of external circumstance. But the cause of life's directionality does not pre-exist or reside outside of the trajectories taken over the course of its unfolding. External causes shape, divert, and constrain it, but they do not explain it. This is the idea behind Bergson's claim that it is the 'movement' through which novel forms are generated that 'constitutes the unity of the organized world', and that the exterior force of 'adaptation explains the sinuosities of the movement of evolution, but not its general directions, still less the movement itself' (CE 105 and 102). Bergson offers the image of 'the wind at the street-corner', dividing 'into diverging currents which are all one and the same gust' (CE 51; cf. Cunningham 1914: 649–50). The air owes a bifurcation in its current to the influence of its encounter with the corner, but the directionality of its movement, which both precedes and survives its division around the corner, has to be explained in another way. No matter how many times it is divided and diverted, each new current continues an original gust in a new direction. It is the unity of the evolutionary movement, by analogy, that each organic form has in common. Just as their unity lies behind them, so does the finalistic force that accounts for that unity, the originary impulsion that is prolonged through the movement that differentiates life forms.

c. Psychology

The immanent finalism of life is virtual, processual, directional, and impelled. Bergson refers to the originary impulsion from which the directionality is derived as the 'indivisible motor principle' of evolution (CE 101). It is the so-called 'psychological cause' of orthogenesis (CE 86). It furnishes what Bergson calls 'the psychological interpretation' of life (CE 51). The originary impulsion and motor principle is virtual tendency. The *élan vital* is Bergson's image for it. That the *élan vital* is an image should go some way towards disarticulating Bergson from the legacy of neo-vitalism. Driesch's entelechy was not an image, but a force, something that exists but is not material. Bergson's *élan vital* is an image, not a force, and is not anything that exists. He considers the image appropriate because 'no image borrowed from the physical world can give more nearly the idea of [the unity of life]' (CE 257).

The 'essence' of psychological phenomena 'is to enfold a confused plurality of interpenetrating terms' (CE 257). When Bergson refers to psychology, he usually has the idea of interpenetration in mind. Physical images reduce interpenetration to the spatialised combination of *partes extra partes*. When Bergson says that 'life is of the psychological order', he means that the idea of interpenetration is essential to understanding it (CE 257). The unity of life is a psychological unity, constituted by a plurality of interpenetrating terms. What are the interpenetrating terms that comprise the unity of life? On the individual scale, interpenetration is a matter of an organism's developmental continuity. On the evolutionary scale, interpenetration is a matter of tendencies.

Evolutionary tendencies are like 'psychic states, each of which, although it is itself to begin with, yet partakes of others, and so virtually includes in itself the whole personality to which it belongs' (CE 118). Mental states are both distinct, differentiated from each other, and co-implicated or continuous. A person's psychology suffuses all of their mental states. Mental states can be considered independently, but only by abstracting them from their inclusion in a total personality. The mental 'whole' is virtually instantiated in each of its states. That is part of what it means to say that the states interpenetrate. The unity of life is a unity of divergent tendencies. Like distinct mental states, these tendencies interpenetrate, retaining their relation to each other and to the whole from which they are dissociated. They are continuous or co-implicated. But a personality is not just a collection of particular states. It is also characterised by an overarching tendency of its own, a tendency to particularise itself in those states. Each state follows from the whole in some direction, realising the tendency of the personality in some form. When Bergson says that 'life is tendency', he means tendency in this global sense as well (CE 99). Like the unity of a personality, the unity of life is characterised by a tendency of its own. This tendency, the tendency of life as such, is a tendency to particularise itself, dissociating and dividing a qualitative multiplicity into particular tendencies over time (CE 57).

Evolution, the movement that unifies all organisms with each other, is a single, still-happening event. It is directed by the materialisation of tendencies over time. Those tendencies are each enfolded in an interpenetrating whole, defined by a tendency of its own. That tendency is the originary impulsion of Bergson's finalism. It is what Bergson refers to as psychological, *élan vital*.[4]

One of *Creative Evolution*'s central definitions of life characterises evolution as 'consciousness launched into matter' (CE 181, 261).[5] This definition lacks the qualification that psychology (or in this case, consciousness) provides what is only an image for life. What Bergson says is that it appears 'as if a broad current of consciousness had penetrated matter' (CE 181). The 'as if' is important. In referring to 'consciousness',

Bergson again has interpenetration in mind. Consciousness is a 'current' – elsewhere he says 'wave' (CE 250) – 'loaded, as all consciousness is, with an enormous multiplicity of interwoven potentialities', that is, interpenetrating tendencies (CE 181). What makes the evolutionary movement look like a current of consciousness is this multiplicity of interpenetrating tendencies. In the equation of life with consciousness, it is the idea of interpenetrating tendencies that is at stake. Bergson is not contending that consciousness really did penetrate matter in the constitution of the first living cells, or that consciousness + matter = life. He says rather that it is *as if* that were the case.

Evolution looks like consciousness because evolution is open-ended movement; and movement or mobility is an indivisibly continuous, essentially temporal phenomenon (CE 128, 155). The idea of stable objects traversing inert spatial locations is an abstraction (CM 125). In reality, everything is moving and changing at different rhythms and rates. Movement and change, properly understood, are synonyms for time. Each refers to the qualitative, indivisible continuity of a process, a trajectory.

> Movements, regarded in themselves, are indivisibles which occupy duration, involve a before and an after, and link together the successive moments of time by a thread of variable quality which cannot be without some likeness to the continuity of our own consciousness. (MM 268)

Motion is a continuous phenomenon. Any point in the evolving trajectory of a movement retains its history of development and tends forward, towards a variable limit (CE 162–3). Each phase interpenetrates in a movement in the process of advancing. Motion or change is defined by the same protention-retention structure that consciousness exemplifies. When Bergson claims that 'between mobility and consciousness there is an obvious relationship', or that 'every reality has a kinship, an analogy, in short a relation with consciousness', he has retention, protention, interpenetrating continuity, and tendency in mind (CE 109; MM 304–5). Conceived in terms of the defining characteristics of movement, life can be seen to evolve 'exactly like consciousness, exactly like memory' (CE 167; cf. Tellier 2008: 428–42 and Bernet 2010). Consciousness, memory, and life are all defined by 'continuity of change, preservation of the past in the present, real duration' (CE 23). The evolution of life is like the evolution of consciousness in these respects. Our reflective awareness of the nature of consciousness provides a principle of access, via analogy, to the nature of evolution (CE 54, 201–2).

There are five other mentions of life's psychological nature in *Creative Evolution*:

- . . . evolution must admit of a psychological interpretation. . . (CE 51)
- . . . if the essential causes working along these diverse roads are of psychological nature, they must keep something in common in spite of the divergence of their effects. . . (CE 54)
- Neo-Lamarckism is therefore, of all the later forms of evolutionism, the only one capable of admitting an internal and psychological principle of development . . . But the question remains whether the term 'effort' must not then be taken in a deeper sense, a sense even more psychological than any neo-Lamarckian supposes. (CE 77)
- We have tried to prove . . . by the example of the eye, that if there is 'orthogenesis' here, a psychological cause intervenes. (CE 86)
- All that which seems *positive* to the physicist and to the geometrician would become, from this new point of view, an interruption of inversion of the true positivity, which would have to be defined in psychological terms. (CE 208)

It is too easy to misconstrue the psychological language (see Ansell-Pearson 2002: 137 and 2005b: 68). Bergson could probably have done more to qualify these claims.[6] They should be read as heuristics, as images that afford epistemic purchase on a reality that outstrips them. Life is not essentially psychological, but the comprehension of some of life's defining features may benefit from the use of psychological terms (CE 257). While we come closest to the truth of evolution when we grasp it as a kind of consciousness, it is nevertheless only 'for want of a better word' that consciousness is what we call it (CE 237, 186).

The heuristic status is clearest in Bergson's contention that 'evolution must admit of a psychological interpretation which is, from our point of view, the best explanation; but this explanation has neither value nor even significance except retrospectively' (CE 51 tm). The effort or impetus at work in psychological life consists in the externalisation, and actualisation, of distinct parts from out of their mutual interpenetration as virtual tendencies. The psychological interpretation of evolution understands the evolutionary movement to be driven by an impetus to externalise interpenetrating tendencies and develop them as far as they can go in actual material form (see Carvalho 2008; cf. Carvalho and Neves 2010: 640). This interpretation is supposed to be the best we have, and yet, since it operates on the basis of an analogy with our consciousness, it is only ever retrospective. It is only after some biological event has assumed a determinate shape for us that we can grasp the impetus at work throughout its process of formation on the model of the tendencies expressed through our own creative activity in time.

Conclusion

The unity of life as a movement is a matter of the common origin shared by every evolutionary trajectory. There are two senses to this claim. According to the first, the evolutionary movement is unified historically, as each lineage originates in a common ancestor, something like the way every marble sculpture human history has ever produced can be traced back to the biogenic formation of minerals. Every distinct species and individual is a moment in the elaboration of a single history, an unfinished event, and can be traced back to the one common origin of all evolvable life. Its difference from the history of marble is that every present living being retains the whole evolutionary past in the form of organic memory. This is what it means to say that life endures (CE 15). According to the second sense of the claim, the common origin of all life is the manifestation of a densely heterogeneous tendency towards indetermination, implicating an interpenetrating mass of virtual tendencies as well (CE 258). This global whole is contracted and instantiated in the actuality of each evolutionary trajectory, unifying every living being with every other as parts dissociated from the whole that implicates them.

The *élan vital* is an image for the originating condition of all living forms as well as the immanent unification of the divergent directions taken by the evolution of those forms via the register of virtual tendency that is instantiated across each. The position that all organisms are unified in a directional movement is better understood as a form finalism than of vitalism. This finalism consists in a reconception of externalism, an inversion of its location of unity, and a reconfiguration of what it is that accounts for the existence of that unity in the first place. Bergson conceives the 'whole' to which purposiveness is attributed not as the sum total of living entities in harmonious relation with each other, but as the evolutionary movement whose unfolding generates each of those entities from out of itself. As I will discuss in the next chapter, stable forms – whether of the organ, the individual, the species, or otherwise – are best understood as the outlines of the directional movements that run through them. Bergson detaches externalism's principle of unification from the postulate of a pre-existent end and relocates it in the originary impetus behind the evolutionary movement. He reconfigures the commonality of the origin of movement according to the psychological interpenetration of life by reformatting it through his modal mereology and positing it as both the register of interpenetrating tendencies as well tendency's essence, which is to extend itself to its limit, dividing in response to obstacles in order to extend itself further in divergent directions.

Bergson's reformulation of finalism is intended to account for the orthogenetic nature of evolution without reducing directionality to a composition of mechanical causes on the one hand or having to posit a set of pre-existent ends on the other. 'We have tried to prove, on the contrary', he writes, 'that if there is "orthogenesis" here [in instances of evolutionary convergence, for example], a psychological cause intervenes' (CE 86). The 'psychology' of this cause entails its determination in the form of virtual tendency. In the next chapter, I examine the set of images through which Bergson presents the relationship of this psychological cause to the stable forms through which it pursues its orthogenetic progress. These images – the canal and the hand – depict the immanence of directionality's cause to the evolutionary processes that it directs.

6

Canalisation and Convergence

Convergent evolution is not an accident or anomaly of an otherwise random variational process. What it reveals is the deep unity that underlies the contingent plurality of life forms, and a kind of pattern, or orthogenetic trend, to their repeated convergence on the same organs and traits. Convergence is best explained in terms of the canalisation of organs by their functions. Canalisation is an image for how one indivisible movement – the tendency towards the realisation of a function – manifests itself in a series of organic structures. Evolution converges on the eye across a series of lineages because each lineage canalises the same tendency towards vision.

Introduction

This chapter concludes the book by situating two of *Creative Evolution*'s most important topics – canalisation and convergence – within Bergson's psychological interpretation of orthogenesis. I show that organic forms are best understood as 'canalising' virtual tendencies through a diversity of material structures. This account is left undeveloped in Bergson's images of the canal and the hand thrust through iron filings. I reconstruct his position on convergence, or the appearance of like forms across divergent lineages, on the basis of those images. Convergence testifies to the virtual unity of life by revealing what is common beneath the proliferation of diverse forms. What is common is a shared starting point as well as the unified directionality of virtual tendencies. As the same tendencies develop through divergent channels, they tend to materialise like forms as solutions to like problems. The camera eye is the most famous example. Its centrality to Bergson's account should be read in terms of the relation between vision and indetermination. The idea that there is a tendency towards the development of vision realised across a number of distinct lineages does not require the postulation of an end that would pre-exist its material instantiations. The idea of open-ended directionality is key to the orthogenetic conception of life as an evolutionary movement trended by virtual tendencies and channelled through processes of speciation.

Canalisation

In Chapter 4, I introduced the distinction between process and product in the context of Bergson's contrast between the organisational simplicity and structural complexity of the camera-like eye. The process/product distinction can be transposed onto the evolutionary register. The materiality of the individual organism is the spatialised 'outline of a movement', an evolutionary movement (CE 128). The forms of individual organs are a kind of silhouette around the progress taken towards the realisation of their functions (CE 95). The relationship between stable forms and the movements that they temporarily stabilise is an index of the relationship between the evolutionary and individual registers of life. They are two ways of determining the same unfolding event: the individual as product, the evolutionary as process. This idea is what Bergson's images of the canal and the hand depict.

a. Images for development

Bergson introduces the image of the digging of a canal only briefly before appearing to replace it with his better-known image of the hand thrust through iron filings, but the two images work in tandem and cannot be treated independently of each other. The image of the hand serves to extend the illustration of the concept of canalisation from the image of the canal. There is one concept – that of canalisation – and two images for it.

The canal is an image for the idea that 'the vision of a living being is an effective vision, limited to objects on which the being can act' (CE 93). In *Matter and Memory*, Bergson theorised the way living beings select a sphere of perception-images on the basis of their sensorimotor capacities as an operation by which a material field is narrowed into a navigable world of experience. In *Creative Evolution*, he describes the pragmatically limited capacity to see as 'a vision that is *canalized*' (CE 93). 'Vision', he claims, 'is a power which should reach *in principle* an infinity of things inaccessible to our eyes' (CE 93). Such a vision would be impractical for the vital necessities of living organisms. A vision that is canalised is a vision that is trained on the objects that are relevant for it, in view of the ability to act on them. Canalisation denotes the tapering down of a possibility space for some effective purpose.

Before referring the 'canalisation' of 'effective vision' to the image of the digging of a canal, Bergson adds that 'the visual apparatus simply symbolizes the work of canalization' (CE 93 tm). By 'visual apparatus' he means the structural complexity of the organ. Vision is canalised according to what is relevant for it. The visual apparatus is a product (or symbol) of the work or activity of that canalisation. Far from generating the function of vision as a capacity of its structural organisation, the visual apparatus is supposed to be an effect of the function instead.

Bergson introduces the image of the canal in order to elucidate the nature of this inverted relationship.

> The creation of the visual apparatus is no more explained by the assembling of its anatomic elements than the digging of a canal could be explained by the heaping-up of the earth which might have formed its banks. A mechanistic theory would maintain that the earth had been brought cart-load by cart-load; finalism would add that it had not been dumped down at random, that the carters had followed a plan. But both theories would be mistaken, for the canal has been made in another way. (CE 93–4)

This is all Bergson has to say about the image (see Al-Saji 2010: 152–3). He turns immediately afterward to the hand, submitting that 'the process by which nature constructs an eye' – presumably analogous to that by which 'the canal has been made' – can be fixed 'with greater precision' in the image of a hand thrust through iron filings (CE 94). It may seem tempting to leave the canal to one side and focus upon the image of the hand instead. This common interpretive move is a mistake on four counts. First, the term 'canalisation' was introduced in order to explain the relationship between the function of vision and the structure of the visual apparatus before the image of the canal was produced and displaced (CE 93). Second, the canal insists throughout the description of the hand, whose movement is said to be *canalised* by the iron filings that outline it (CE 95). Third, the term reappears in later, pivotal discussions of the thermodynamic accumulation and release of free energy in the living being (CE 110, 126, 256). And fourth, there exists a line running from Bergson's use of the term through Whitehead's process metaphysics to the embryological theory of C. H. Waddington and from there into mainstream embryology today (see Helm 1985: 154; Smith 2004: 123; Morris 1991: 42). Though these connections have so far been overlooked in the scholarship (see Al-Saji 2010: 171, n. 10), I think there is something to be learned by pointing them out.

Bergson describes a canal whose construction is not adequately explained by the accumulated dirt that makes up its banks. Canals are formed in three ways. (1) A channel is dug into dry land through excavation and supplied with water from elsewhere. (2) A channel is dredged into the bottom of an already existing body of water that is subsequently drained down to the canal. (3) Raised banks are constructed in parallel in order to create a channel between them by outlining it. As Bergson mentions 'digging' and the 'heaping-up' of earth in banks, we might imagine a canal of the first and the third kind. This canal seems to have had its channel dug out and its banks formed out of the dirt displaced. A water supply would flow through the negative space of the channel and adopt a trajectory contoured by the material structure that directs it. The flowing water stands in relation to the structure of the canal in a manner analogous to the relation between the banks and the channel. Whereas the negative space of the channel corresponds to the visible structure of the floor and banks that define it, the canalised flow corresponds to the static material composition of the canal that defines it.

The process through which this kind of canal is formed stands in the same relation to the constituted material of its structure as both the negative space of its channel and its banks stand to the flow of its water and the material delimitations that contour it. Dirt was dug out from a stretch of land to form the canal's floor and piled alongside it to form a set of parallel banks. The finished banks signify the work that went into accumulating the soil for them. The digging of the floor is also what establishes the trajectory of the canal. The banks are assembled around it. While their formation completes the structure of the canal and allows it to channel a flow, the canal's banks are derivative components in its formation, both in terms of their material as well as of the trajectory of the channel that they delimit.

The canal is an image for a fully developed biological structure. Bergson thinks that 'mechanism would maintain that the earth had been brought cart-load by cart-load; finalism would add that it had not been dumped down at random, that the carters had followed a plan' (CE 93–4). Both would agree in attending only to the finished product, missing the process responsible for its formation. While that process – the digging out of the canal floor – consists primarily in the generation of the negative space of the channel, its only visible material effects are embodied in the floor and banks around it. Mechanist and finalist explanations both privilege the banks, or finished parts, disagreeing only over whether they were constituted accidentally or by plan, because both positions mistake what are the visible products of an invisible process of formation for its real component elements.

The difference between the activity of excavating dirt and its accumulation in banks can be applied to the difference between the material constitution of those banks and the flow of water canalised by them as well. In both cases there is a difference between a temporal process and its static material correlate. In the case of the canal's formation, the static character of the latter has to be explained via the processual nature of the former. Materially conceived, the canal provides the structural conditions for the channelling of a flow of water, but conceived as a process, the canal works by drawing water from a larger reserve, concentrating it in order to direct its flow through a channel, and guiding its trajectory elsewhere for some purpose. The sedimented materiality of the canal figures both as the material product of the process of its formation and as the material substrate for canalising movement. Thus, the image of the canal depicts the distinction between process (formation) and product (material structure), illustrating the irreducibility of the former to the latter as well as the reincorporation of product (the canal) into process in turn (canalisation).

After asserting that both the canal and the eye have been made in a way that eludes the explanatory resources of mechanism and finalism alike, Bergson replaces the image of the canal with the image of a hand and, after elaborating the details of the second image, returns to his analysis of the structure and function of the eye. He introduces this image by adding the idea of resistance to the example of raising one's hand (see Fujita 2007). 'Let us now imagine', he says, 'that, instead of moving in air, the hand has to pass through iron filings which are compressed and offer resistance to it in proportion as it goes forward' (CE 94). By offering resistance, the filings 'canalise' the trajectory of the hand by limiting its motion and exhausting the effort required to push any further. Bergson imagines that the outline of the hand remains preserved in the arrangement of filings wherever it stops, and that the hand and arm are invisible. The result is a mass of iron filings coordinated around the silhouette of an absent hand. The image is analogous to the canal in the bifurcation between its material configuration and the invisible process responsible for its formation.

The coordinated filings should not be thought as a causally inert representation of the hand's final position. The reason why the image involves filings and not a more plastic medium is because the filings resist the hand that is plunged into them, canalising its movement and exerting a causal influence on its trajectory. The filings can be conceived both as the product of a canalised process and as the material substrate for the canalisation of that process. The distinction between process and product is at least partially bidirectional. The filings are a spatialised expression of the trajectory taken through them as well as ingredient factors in the calibration of that trajectory itself.

b. Vision and its apparatus

Bergson proposes that 'the relation of vision to the visual apparatus would be [on his hypothesis] very nearly that of the hand to the iron

filings that follow, canalise and limit its motion' (CE 95). Vision is canalised by the physiology of the visual system. The structure of the organ,
as the material substrate for the canalisation of vision, is what explains
the function's pragmatic efficacy. An organism sees in its environment
only what is meaningful for it as an effect of its body's material constraints on what it can do. In this sense, 'one could say', as Jankélévitch
does, 'that the animal sees *despite* its eyes rather than by means of them'
(2015: 138). It is a strange way of putting the point. The idea is that the
body is as much a limitation on the functional capacities of its organs as
it is the instrument of them. Limitation is facilitation's material means.

If vision is canalised by the visual apparatus, then any attempt to
explain the composition of the apparatus on the basis of its structure
alone would be akin to mistaking the constituted banks of a canal for the
process through which the canal was dug. Structures are the products of
developmental processes. They are not primary over their functions and
do not explain the functions that correspond to them. When Bergson
calls the materiality of the organ 'a negation rather than a positive reality',
he means both of these things. The organ is the embodiment of its function and the spatialised outline of the process of its formation (CE 93).
Organic forms 'counterfeit immobility so well that we treat each of them
as a *thing* rather than as a *progress*, forgetting that the very permanence of
their form is only the outline of a movement' (CE 128).

The functional integration of the visual apparatus is what explains the
composition of its structural elements. This idea has an affinity with 'downward causation' approaches to biological explanation (see Campbell 1974).
Philosopher of science Sara Green observes, for instance, that the functional capacities of an organ cannot be explained by the physical composition of its structure, since the function does not obtain at the micro-level of
the structure's parts (2007). This is sometimes called 'medium downward
causation'. Bergson holds something like this position, but goes further,
since for him it is the function that is responsible for the composition of the
corresponding structure. The function exerts downward causation on the
organisation of the organ in which it is realised.

On Bergson's view, the organic structure is a medium for the
expression of the function. The greater the hand's effort, according
to his image, the further its trajectory through the filings that resist it.
The visual organ's sophistication – its degree of complexity, the coordination among its parts – stands in proportion to the advancement of
'the undivided act constituting vision' (CE 95). The complexity of the
biological structure is a measure of functional intensity, not the cause
of it. That function is embodied to different degrees of intensity in different visual organs. Behind all of the visual apparatuses of the animal

kingdom – from gastropods and spiders to cephalopods and mammals – there is just one function, one tendency, one movement.

The formation of individual structures is insufficiently explained by their embryogenesis alone. The specific form of each individual structure 'only expresses the extent to which the exercise of the [evolutionary] function has been obtained' (CE 96 tm). Embryogenetic processes canalise the evolutionary movement (CE 96). The tendency towards vision is not inhibited by an opposing tendency, but by the material substrates that realise it. Vision is exactly as functionally intense as organic structure allows. Just like the filings, 'the materiality of the organ [of the eye] is made of a more or less considerable number of mutually coordinated elements', in direct proportion to how far vision was able to realise itself in that medium (CE 95). The instantiation of vision in the eyes of the animal kingdom is like the plunging hand spatialised at various stages of movement.

The evolution of vision, in the form of organs like the camera-like eye, plays itself out across the history of life as a conflict between the tendency towards a maximum of visibility and the limits placed on it by the material apparatuses that condition its realisation. The camera-like eye cannot appear in a species whose physiology is inadequate to the construction of an iris, retina, and vitreous cavity or whose neurology cannot make anything useful of visual information. The progress towards the camera-like apparatus is a correlate of the complexity of its medium. Beyond an impediment to the realisation of vision, the materiality of each visual organ is also an achievement. Every determinate visual structure testifies to the series of structures traversed by the function of vision before realising itself in it. The gastropod eye is a limit with respect to the abstract future towards which its function tends, but a progress with respect to the past consolidated in it. This is true of every organ. Canalised and canalising, it is at one and the same time a limit on and an achievement of the corresponding function.

The appearance of visual organs, eyes, as complex structures distinct from the visual function that they realise may be misleading. 'The eye with its marvelous complexity of structure, may be only the simple act of vision, divided *for us* into a mosaic of cells' (CE 90). Structural complexity is a human contribution to what is in nature a 'simple act', no more complex than the lifting of a hand (CE 91). The difference between the simplicity of the process and the complexity of the product that outlines it is an effect of perspective. 'Vision is nothing', to borrow again from Jankélévitch, 'but the extension and as if the blossoming of the visual apparatus' (2015: 118). Seeing and eyes are two words but not two things. We treat vision from the point of view of manufacture, likening the eye to the structure of a built artifact and vision to its function and supposing them separate (CE 93). We can build many different kinds of engines, for instance, realising the same function across a variety of material structures. But no such separability obtains in evolution, which does not proceed by trying to generate a function on the basis of a structure as if it had the former in mind while assembling the latter.

Bergson's position should not be understood as a dispute with the idea that structures can be repurposed for new functions, such as the fish swimbladder in Darwin's famous example (see Bennett and Posteraro 2019: 4–5). Since structures are integrated with their functions, a change in one implies a change in the other. When a structure is used for a new form of activity, it is a new kind of thing. This does not mean that such changes cannot or do not occur, but only that when they do, they do not testify to the blind indifference of a mechanical structure to its function. They are implicated in a general shift in the integrated unit as a whole, perhaps even the organism of which they are a part as well.

'Function is', in organisms, 'an intimate and necessary continuation of the structure', analytically separable but organically indissociable (Jankélévitch 2015: 118–19). We take one process, that of seeing, and divide it off from the function of sight, which we consider to be realisable in any other appropriately formatted substratum. What we have left is the visual structure, which we liken to a machine and decompose into the complicated assembly of cells, muscles, membranes, and tissues that we discern within it. But these are each images, and their discontinuity is an artifact of vision, a perceptual selection performed on what is a simple whole. The idea that they are related to each other as parts in a machine is an artifact of analysis, a misplaced introduction of the utility of manufacture into the domain of life. 'Nature', says Bergson, 'has had no more trouble in making an eye than I have in lifting my hand' (CE 91). The evolutionary process does not have to know the anatomy of the eye in advance before tending in the direction of visual perception.

c. The inside of indetermination

By positing a tendency towards the realisation of one function across organs of varying complexity, Bergson may seem to ascribe teleology to the evolutionary process. That is not exactly right. Vision is not an end laid out ahead of the evolutionary processes that approximate it (CE 96). The function does not exist for species that have not already obtained it. If evolution does evidence a progress towards vision, it is not on account of the existence of an end possible before being actualised. Vision should be understood as an immanent tendency, a natural result of the evolutionary movement. Bergson says that 'it is really effected in virtue of the original impetus of life; it is implied in the movement itself' (CE 96). We should expect to see it manifest wherever materially possible, and just so far as materially possible, just as we should expect to see metabolism or motility appear wherever their conditions are secured as well.

Vision is immanent to evolution because vision qualifies the subjective facet of the primary function of life, to intensify indetermination. It consists in the sensory presentation of choice to motile organisms. I mentioned Bergson's observation that 'between mobility and consciousness there is an obvious relationship' (CE 109). In the previous chapter I discussed that relationship in terms of duration, the retention of the past, and the interpenetration of tendencies in a qualitative multiplicity. But there is another dimension: the 'power to move *freely*' (CE 111). The ability to move freely is a developmental eventuality, along one line, of the effort to maximise and utilise indetermination. As the animal nervous system evolves, 'the more numerous and more precise become the movements among which it can choose; the clearer, also, is the consciousness that accompanies them' (CE 110). Consciousness is a correlate of adaptive sensorimotor ability, presiding over action and enlightening choice (MM 182). For a body to navigate its environment successfully, it has to orient itself with respect to its surroundings. It has to model the environment as a navigable terrain and itself as a set of locomotive abilities within it. A being unconscious of itself within its environment is incapable of free movement.

Consciousness, in this basic iteration, is another name for the set of facilitating conditions for active movement (MM 46). Movement is free in proportion to the choices available to it. Bergson defines choice as 'the anticipatory idea of several possible actions' (CE 96). The animal's ability to choose among possible actions makes of its environment a set of contingent possibilities. 'Possibilities of action must therefore be marked out for the living being before the action itself' (CE 96). Possibilities of action are virtually instantiated in the animal's perceptual world. Perception is virtual action, and 'visual perception is nothing else; the visible outlines of bodies are the design of our eventual action on them' (CE 96).

Perception, vision included, is defined by the tendency towards action. It is completed when some behavioural suite of actions is set into motion, actualising what in perception was virtual. Virtual action – the perceptual presentation of possibilities – is what choice looks like for an active, moving body. Immanent to the tendency towards indetermination is a corresponding tendency towards vision as the perceptual mechanism by which the organism marks out choices in its perceptual milieu.

If there is a tendency towards vision, it is because the tendency towards indetermination includes a sensory, subjective dimension, an inside, and vision is a quality of the appearance of indetermination to consciousness. The inside of indetermination is formed through the reception of external stimuli, its conversion into information about what the animal can do in its environment, and the instantiation of that information in the form of virtual actions in the images of perception. The visible manifestation of those virtual actions comprises only one of the many facets of subjective indetermination. Bergson seems, then, to accord a special, perhaps even undue privilege to vision (see Lawlor 2003: 5–6; cf. Mullarkey 2004: 482). Any of the other perceptual capacities involved in the animal's sensorimotor navigation of its environment would seem to fulfil the criteria supposed to set vision apart. That it is vision alone that is made to represent the living experience of a world in terms of possibilities for action appears to be unjustified.

Ansell-Pearson argues that the privilege is accorded to hearing instead, which alone discloses a continuity of experience (2002: 12). Mark Antliff makes the same argument, referring to Bergson's use of the example of melody as an image for duration (1991: 171). I think that this is incorrect, both as a description of auditory sensation, which is also trained by attention on virtual action, and as a denial of the way Bergson valorises vision. There are a number of reasons to motivate the privilege accorded to vision as a consequence of the special affinity between Bergson's use of the concept of image and the place of art in his thought (see Lawlor 2003: 7–10). I do not think these suffice to justify the privilege either. Every sensory modality has its corresponding artistic practice, and Bergson refers, in addition to the visual arts, to literature and music as well (see Lorand 1999; cf. Sinclair 2020: 177–99). I do not think it is a stretch to speak of poetic and auditory images, and perhaps even tactile and gustatory images (with respect to the culinary arts, for instance). At the very least, these observations should complicate what might appear to be a straightforward identification of the visible image with the artistic one.

I consider vision to be a synecdoche for the subjectivisation of indetermination. Vision and its apparatus might be seen, on this suggestion, to stand in for the total set of the body's sensory capacities and the complete sensorimotor system in which they are embedded. Not only are the criteria that make vision a part of the subjectivisation of indetermination fulfilled by other sensory capacities, but a given animal's capacity for sight is also continuous with the rest of its sensorimotor system. Useful sight is far from a single operation. Receptivity to light, at least in locomotive animals, means being able to navigate a light-saturated environment. 'Our eye makes use of light', as Bergson puts the point, 'in that it enables us to utilize, by movements of reaction, the objects we see to be advantageous, and to avoid those which we see to be injurious' (CE 71). That requires a prolongation of the retina 'in an optic nerve, which, again, is continued by cerebral centres connected with motor mechanisms' (CE 71). To speak of the visual apparatus is to 'allude to the very precise relations that exist between this organ and the apparatus of locomotion' (CE 71). Vision is possible only by means of the other systems – nervous, muscular, osseous – that are 'continuous with the apparatus of vision in vertebrate animals' (CE 71). Any of the claims made about vision are at least partly applicable to the rest of the body's sensorimotor system. One might just as well say that the auditory faculty is evolved just so far as the material complexity of its sensory system allows. If vision is equated with the inside of indetermination, then it should be as a stand-in for the total set of sensorimotor functions with which it is continuous.

Vision should not be privileged as a particular function, correlated with particular eyes. If vision does mark a difference, it can be only in degree, not kind. It does seem to mark such a difference of degree, as vision appears to furnish the living being with a reservoir of perceptual images wider, deeper, and more intricately articulated than those of other sensory modalities. The perceptual content that vision delivers has the potential to outstrip the otherwise rigorously practical orientation of the rest of perception. Its 'effect', as Bergson observes, 'goes beyond its object' (TS 144).

Bergson refers twice to the fact that 'we can see the stars, while we have no control over them' in order to make the point (TS 144, 222). This may be an uncharacteristically careless formulation. The fact that we lack control over the stars does not count against the practical nature of the perceptual images that we form of them. Not every object of perception is an object on which we can act at the moment that we perceive it. We perceive in matter only what our body is capable of acting upon in general. We form the images that we do because of an evolved, adaptive benefit of forming images in that way and not others, at a certain scale and gradient. Looking skyward, we register stimuli from incoming light waves and filter it as we do all other stimuli, according to the exigencies of our interactions with material bodies. We do not see stars *as they are*, whatever that would mean; we see them according to the way our visual apparatuses have evolved to generate perceptual images at a scale on which we can act (see Weiss 1952).

Can we interact with stars? Not if action requires physical manipulation, but why should it? If that is the standard, then most of what we perceive eludes our capacity for interaction. Bergson later clarifies the suggestion that we can perceive in excess of what we can act upon by distinguishing within the category of 'interaction' immediately actionable possibilities from what he calls 'theoretically possible actions' (TS 222). In the latter class we should place the work of Galileo, Kepler, and Copernicus. What is astronomy if it is not composed at least in part of interactions with the perceptions of celestial bodies? Vision gives rise to an effect that goes beyond its object because vision gives rise not only to those perceptual images with which we can interact physically and immediately, but to those images with which we can interact theoretically as well (TS 144). By supplementing this form of action with an array of tools and techniques, we can learn to perceive matter on a scale that would otherwise elude us. Either these perceptions are communicated to us by our technologies in the form of data, or else it is vision that is doing the perceiving, aided by imaging technologies but able to see nonetheless.[1]

In addition to being the furthest reaching, vision's images are also the best articulated and most discontinuous. Vision delivers a perceptual field populated by mid-range objects distinct from each other and set against the background of their environments (MM 278). This is largely an effect of scale, as what look to us like distinct material bodies are the perceptual effects of a smaller scale field of fluctuating vibratory activity (MM 266). It is also an effect of the evolution of vision as a preparation for touch (see Ansell-Pearson 2002: 12). We see so that we can handle, grab, consume, combat, and use (see Turvey 1992). For those purposes we require the images of distinct things. The other sensory modalities seem less tactile-oriented, less trained on physical manipulation. Auditory phenomena are not rigidly demarcated from each other or as individuated into separate bodies. They are continuous, differentiated spatially and by means of the shifting dictates of attention and concern, but not in their presentation to sense awareness the way visual images are. The same appears to be true of the gustatory and olfactory senses. This is obviously the case for the proprioceptive and vestibular senses too.

On account of their discontinuous articulation, visual images are especially receptive to the instantiation of virtual actions (see Gibson 1979: 127). They suggest what can be done with them the most clearly (CE 96). Visual images may afford the most possibilities for action. They may present organisms with an especially wide range of available responses, navigatory directions, and provisional ends for the schematisation of future activity. Perhaps visual images are best suited as facilitating conditions for an organism's ability to move and act freely. It would seem to follow that consciousness, as a correlate of self-movement, is at its most intense − at least in principle − in the extrapolation of vision to its limit. The sensation of choice, as the inside of the tendency towards indetermination, would seem to tend towards vision as if towards its own limit. Vision would seem to represent the fullest realisation of sensation of which we know.

If there is an objective tendency towards indetermination, then there is a corresponding subjective tendency towards vision as a result. If there is a privilege to be accorded to vision, then it is in this sense. Vision can be considered the abstract end defining the development of virtual actions as the subjective face of indetermination, abstracted away from the total sensory system in which it is embedded and extrapolated to the limit of the power proper to free movement.

By conceiving the function of vision in a manner that corresponds to the tendency towards indetermination, by locating vision internal to and partly coterminous with the evolutionary movement, Bergson is able to postulate it as a tendency whose completion orients the evolution of visual apparatuses across the animal kingdom without thereby lapsing into the finalism he already criticised. If there is a directional constraint on the evolution of motile animals in its favour, it is not because there exists a specific visual apparatus as evolution's goal, but because evolution trends in the direction of indetermination among moving organisms and vision is the perceptual capacity best fitted to that fact.

Both vision in particular and indetermination in general are canalised by the structures that facilitate them. In Chapter 4 I showed how the tendency towards indetermination could be understood on the basis of its physiological canalisation through the metabolic processes that afford the animal its latitude of possible actions in response to external stimuli. This tendency is canalised subjectively via the sensorial presentation of choice to consciousness, contoured by the material means by which visual information is captured, processed, and converted into the visual images in which possibilities for action are instantiated. Particular organisms assume their structural configurations contingently, as a result of the way energy is canalised through the metabolic processes as an effect of the composition of the temporalising and spatialising tendencies. Particular visual apparatuses assume their structural configurations in the same contingent fashion, as a result of the way information is canalised through sensorimotor systems and completed in action. This is an effect of the composition of the temporalising and spatialising tendencies as well, in this case of the tendency towards choice in relation to the materiality of the perceptual organs through which that choice is processed and registered in images.

Trends are the results of canalisation. By the end of *Creative Evolution*, it is trendedness that comes to comprise the process's primary sense

(CE 110, 126, 256). With some qualifications, it is this sense that is retained by the use of the term 'canalisation' in the embryology of today (see Salazar-Cuidad 2007). It was most likely Whitehead who first discerned in Bergson's image the value of this aspect in particular (see Helm 1985: 154–5; Urban 1951: 304; and Robinson 2009: 2). The question of the extent to which Bergson played an influential role in the development of Whitehead's philosophy of organism has been the subject of some debate since F. S. C. Northrop declared Bergson the chief resource for 'the basic concept and doctrine of Whitehead's entire philosophical and scientific outlook' (1941: 169). Northrop identified Wildon Carr's 1912 monograph, *Henri Bergson: The Philosophy of Change*, as the most likely intermediary between the two. The critical response to this suggestion is exemplified by Victor Lowe, who argues that Bergson exerted little to no positive influence on Whitehead, seeing the two thinkers as having developed some broadly similar commitments independently (1949: 271, 272, 278). In his more recent study on the topic, Didier Debaise concedes an initial convergence, but argues for a basic difference in the way each develops his system (2009). Neither Lowe nor Debaise addresses the idea of canalisation. None of the other critics of the idea of Bergson's influence on Whitehead do either.

Whitehead introduced the term 'canalisation' in *The Principle of Relativity* to emphasise the productive dimensions of limitation, attributing to Bergson both the word and the insight (1922: 16). Though Whitehead did not initially elaborate its details, the term reappeared throughout *Process and Reality*, serving as an essential factor in the explanation of the emergence of order (1978: 107–8, 129, 178). According to Whitehead, canalisation is a positive process. The material of a canalised process should not be understood as the process's spatialised outline, but as the means by which it was progressively tapered, becoming increasingly irreversible over time. In this way, Whitehead conceived of canalisation as a general metaphysical principle capable of explaining everything from gene expression to the order and continuity required for the maturation of personality.

It might not be surprising that Whitehead saw something powerful in Bergson's image of the canal. What he saw was the insight that creativity can only elaborate itself within the confines of material bodies. Without material 'brakes' on the tendency to differentiate, evolution would be able to produce nothing stable at all, nothing on which natural selection could subsequently act. That would not be a victory won on behalf of vital tendency, but a failure, for undirected variation would not be productive as much as chaotic – all experimentation and no artwork. Whitehead affirms and develops canalisation's generative dimensions. Canalisation supplies order, which is required for creativity and an increase in what Whitehead calls 'intensity'.

Whitehead emphasises the way Bergson utilises the image at the end of *Creative Evolution*, as a mechanism for the concentration of an initially diffuse activity. He recodes the idea of productive concentration in the terms of his own system. In Whitehead's formulation, canalisation refers to the appropriation – or 'prehension' – of an inherited past of consolidated acts and relations. As the organism develops, it integrates more of its relations and transmits those integrations along a temporal line. The organism does not have to decide how to relate to the world because the outlines of its relation are increasingly rigidified for it. This implies a certain degree of irreversibility, just as it implies a developmental tapering in the space of possibilities.

Canalisation affords the organism an increase in order, which is just as necessary for the intensity of experience as originality is. Originality breaks the strictures of the order out of which it emerges and into which it settles, after expanding it. In biological terms, Whitehead might say that canalisation allows the organism to take a relatively stable route through a chaotic set of interior and exterior milieux. It allows for the reliable expression of phenotype, just as it ensures that different organisms of the same species behave in similar ways and so are able to survive in the same niche. Creativity and originality – or mutation, anadaptation – require that reliability as their starting point. 'Thus life is a passage', on Whitehead's account, 'from physical order to pure mental originality, and from pure mental originality to canalized mental originality' (1978: 107–8).

Creativity necessitates some means by which societies of 'actual subjects', or existent organisms, can interact with what has yet to come to pass. This is what Whitehead means by 'mental originality', the prehension of what exists over and above the actual. For Whitehead, that is the conceptual. If prehensions were just physical, related only to their actual pasts, then novelty would be impossible. Creative advance is conditioned because of the creative decisions of subjects capable of prehending the physical past as in some sense other than it was, conceptually – in the indeterminacy afforded it by what Whitehead calls 'eternal objects'. If physical feeling relates to the settled facts of the past, then conceptual feeling concerns the future, the formal possibilities left open by those facts. Every process of actualisation concretises what is initially (or potentially) indeterminate by prehending the conceptual along with the actual, adding another layer of realised actuality to the ongoing creative advance. Unity and continuity are achievements, not starting points. Canalisation supplies the mechanism of their explanation.

C. H. Waddington – a developmental biologist responsible for initiating the conceptual revolution in the life sciences known now as 'systems biology' – elaborated his theory of the epigenetic landscape and its branching pathways of development while reading Whitehead's *Process and Reality*.[2] Waddington drew on Whitehead's reformulation of the image of the canal as the means by which to explain organismic order in the midst of environmental variability. He sought an explanation for the reliable production of similar phenotypes in a variable population scattered across significantly different environments.

Waddington imagined a plane into which several divergent canals have taken shape. What flows through them is not a tendency in the process of realising itself, but – according to a certain secularisation of Bergson – the developing cell instead. Before becoming canalised along one line of development, the cell's fate is plastic. It can follow a number of pathways, each of which is contoured by the interactions of various genes. Once the cell begins to develop along certain trajectories, it gains in expression – what Whitehead called 'intensity' – what it loses in plasticity. Developmental irreversibility is the key to Waddington's formulation of the image. Becoming canalised means falling into a groove, taking one pathway of development at the expense of initially possible others. 'Developmental reactions', in his own words, 'are in general canalized . . . so as to bring about one end result regardless of minor variations in conditions during the course of the reaction' (1941: 563). It was on the basis of Whitehead's redeployment of Bergson's image of the canal that Waddington developed a theory of epigenetics that, as Adam Wilkins has suggested, was nothing short of 'a premature discovery' (1997: 257).

It is important to note that Waddington's image does not correspond to anything concrete. There are no canals in the cell. It is an abstraction with purely heuristic value. Bergson generalised canalisation across evolutionary history, employing it as an image for the way virtual tendencies are channelled through a series of biological structures over evolutionary time. Whitehead made that image into a principle for the irreversible production of novelty on the basis of ordered and concentrated creativity. Waddington encoded the insight in his formulation of cellular development. It remains important to the study of epigenetic stability today.

Convergence

I mentioned convergent evolution in Chapter 1. Convergence consists in the appearance of like forms across divergent lineages independently of a shared history. Bergson's account of the phenomenon unifies his theory of orthogenesis as a reformulated finalism of tendencies with the concept of canalisation.

Convergence is supposed to indicate a canalisation of the same tendencies through different material structures. Canalisation in cases of convergence involves the retention of a virtual whole in each channel of actualisation. This account is supposed to succeed where actualist (and possibilist) theories of evolution are supposed to have failed. Bergson locates in convergent evolution an empirical justification for his reformulated finalism. Here is why.

> The more two lines of evolution diverge, the less probability is there that accidental outer influences or accidental inner variations bring about the construction of the same apparatus upon them, especially if there was no trace of this apparatus at the moment of divergence. But such similarity of the two products would be natural, on the contrary, on a hypothesis like ours. . . (CE 54)

More so than any other scientific datum, convergence is supposed to recommend Bergson's philosophy of tendency, exemplifying that philosophy's empirical viability and appreciable explanatory power (CE 54–5; TS 112, 115–16). Bergson is right to insist on the potentially revolutionary import of convergence. Many scientists now consider its study an opportunity to re-examine and even revise some of the principles of adaptationism (see Morris 2003: 284 and 2015: 4–6; cf. Pearce 2012: 430–1). There may also be something legitimate about the idea that evidence for convergence is also evidence for 'finality, in the special sense in which we understand it', at least 'in a certain aspect', namely directionality and trendedness (CE 54).

Bergson goes too far when he concludes that evidence for convergence is also evidence against adaptationism, and that mechanistic theories of evolution should prove outright 'refutable' in light of enough such evidence as well (CE 54). Though adaptationism has perhaps had to draw upon outside resources in order to accommodate for convergence, it is far from having been refuted. In fact, there is an important sense in which evidence for convergence is also evidence for the efficacy of selection (see Morris 2003: 303–7). Ansell-Pearson concludes that Bergson's argument from convergence to the *élan vital* should be considered a failure (2002: 92–4; cf. Al-Saji 2010: 150). I agree that convergence does not provide evidence sufficient to validate Bergson's reformulated finalism over the alternatives, and so cannot operate as a litmus test for theories of evolution in the way that he assumed. But the motivations for and implications of Bergson's account should not be discounted as a consequence. Neither should the account be relegated to the status of a merely regulative postulate, as Ansell-Pearson proposes, in order to retain its plausibility (2002: 92–3).

The 'psychological interpretation' of evolution does not stand or fall with convergence. This admission does deprive Bergson of the empirical means by which to ratify what he considers to be the insuperable deficiencies of mechanism. But it also allows me to qualify the place occupied by convergence within his theory of life. Convergence is best understood first as a challenge to non-directional theories of evolution. Even if it cannot refute them, it does require that they go beyond themselves in order to explain it. Second, convergence is an indication of how the *élan vital* might help to explain an empirical phenomenon, in a manner that exemplifies the role it plays for Bergson's finalism. Neither needs to be definitive for them both to advance the critical and creative aims of Bergson's philosophy of biology by specifying those aims in terms of an open research agenda.

Here are the three main components of Bergson's account. First is the unity of tendencies in the virtual whole and the complementarity they retain to each other after they have been dissociated and developed. Second is the 'psychological interpretation' of tendency, which allows for an analogy between the recollection of the same memory-image in different perceptions and the canalisation of the same organic form along different evolutionary channels. Third is the bidirectional dynamism of actualisation, which accounts for the probability that the same small set of forms (such as the eye) are canalised wherever materially possible. I present each before considering the affinities they share with some contemporary scientific models.

a. Definitions

Convergence remains an unsettled category of evolutionary phenomena (see Pearce 2012: 430–1; Beatty 1995; Powell 2007 and 2009). Its relationship to adjacent categories, such as parallelism and analogy, has long been the subject of debate around issues of contingency and direction in evolution. The question of how to conceive convergence occurs at the intersection between internalist and externalist accounts of evolutionary agency (see Godfrey-Smith 1996). To locate Bergson's understanding of the category with respect to the dominant alternatives, I begin by defining some of the relevant terminology.

Externalism is the position that locates evolutionary agency, or what Gould calls 'the motor of organic change', in selection pressures external to a random proliferation of biological variations (1977: 2). This is sometimes called panadaptationism, since externalism takes adaptive benefit with respect to environmental problems as the primary reason for the consolidation of heritable traits. Internalism is the position that selection is not the only force driving evolutionary change. Internalism posits constraints on the development of organisms that function to limit and bias the production of variations. These internal constraints shape the pool of variations on which external selection acts, mediating the force of selection and driving evolution from the inside.

Each position has its own explanation for the independent appearance of like traits in divergent lineages (see Powell 2007). Internalism conceives them as instances of 'parallelism'. Externalism conceives them as instances of 'convergence'. Similar traits are 'parallel' in two species if they have been evolved from out of a shared ancestral character. Though the similar traits did not exist in the ancestor, descendant species inherited a developmental potential for those traits that each expressed in parallel after diverging. Selection pressure alone would not explain the resultant similarity, as it would have been derived from a shared origin. Similar traits are considered 'convergent' when they are independently evolved in different species that did not share a recent ancestor.

In cases of convergence, the resultant similarity would be an effect of congruent selection pressures acting to bring about a like solution to like environmental problems in descendants that did not previously have anything relevant in common. Externalism explains the appearance of like forms in different species as instances of convergence, since it posits selection pressure as the sufficient reason for the evolution of heritable traits. Internalism reinterprets their appearance as instances of parallelism in order to credit shared developmental history as an efficacious factor in the evolution of heritable traits.

Some theorists argue that closely related organisms can evolve the same trait via different mechanisms while distantly related organisms can evolve the same trait via the same mechanisms as well (Arendt and Reznick 2008: 27–30). They conclude that the distinction between parallelism and convergence is better replaced with an inclusive concept of convergence that does not discriminate between differences in how shared traits were evolved. I think Trevor Pearce is right to respond that the parallelism–convergence distinction can be retained if it is premised on the difference between the underlying generators responsible for the appearance of like characteristics, instead of taxonomic distances from common ancestors (2012: 439).

There are two dimensions to the process by which similar characteristics arise on independent lineages. There is the manifest trait and the developmental pathway responsible for manifesting it. Parallel evolution does not consist in the inheritance of the same trait in divergent species but in the pathway, or developmental potential, that is expressed in the later evolution of each trait. Gould distinguishes between 'underlying generators' and 'realized structures' (2002: 1078). Parallelism consists in the inheritance of the same 'underlying generators' in divergent species; the traits they have in common are the 'realized structures' of those generators. They are not themselves inherited but conditioned by the generators that are. The generators are called 'homologous', which means that their similarity is due to derivation from a common source. Characteristics whose similarity was independently evolved are called 'homoplastic'. Homology is similarity by shared origin; homoplasty is similarity by other means. Parallelism involves the homology of underlying generators in connection with the homoplasty of realised structures, while convergence eschews homology and posits structural homoplasty as a result of non-homologous underlying generators selected for via external pressures.

Bergson did not anticipate any such specifications. He writes only of the generation of 'the like apparatus, by unlike means, on divergent lines of evolution' (CE 54). By that definition, he seems to intend convergence in the externalist sense, for 'unlike means' imply non-homologous underlying generators. The 'like apparatuses' that are their homoplastic products would seem to have arisen along divergent lines independently of the ancestral qualities shared across them (CE 54–5). Bergson therefore appears to endorse the externalist contention that evolution can arrive at 'like apparatuses', or homoplastic traits, without employing 'like means', or homologous underlying generators, in the process. Yet it cannot be selection pressure, on Bergson's account, that is responsible for their production. He posits a positive 'motor of organic change' that makes the appearance of convergent forms a natural bias of a trended evolutionary movement. 'Such similarity of the two products [of convergence] would be natural', he writes, for 'one should find even in the last rivulets [or evolutionary lineages] something of the impulsion received at the source' (CE 54 tm).

Bergson's invocation of a shared impulsion seems to correspond to internalism's idea of constraints on the parallel development of homologous pathways. Bergson's 'motor of organic change' is, of course, the virtual impulsion of vital tendency. It is common across all of life because it is external to each ostensibly individuated living form. It cannot be equated with internalism's underlying generators. It is not inherited at any particular point in evolutionary history, derived from any particular ancestral quality. It is beyond the genome of any already constituted species, residing in the movement through which each species is constituted.

Bergson is probably best understood as accepting aspects of internalism and externalism while ultimately frustrating the category requirements of both.

b. Unity and complementarity

The virtual unity of life is composed of a qualitative multiplicity of interpenetrating tendencies. This multiplicity develops through a series of bifurcations, each of which yields an increasingly specialised set of divergent tendencies. Bifurcations are a response to material limitation. Frustrated, a tendency divides instead of culminating. Its elements are apportioned across two new tendencies as a result, each of which develops in a new direction of its own. Tendencies extend themselves further by splitting themselves up.

Evolution unfolds accordingly. 'From its origin, [life] is the continuation of one and the same impetus, divided into divergent lines of evolution' (CE 53). This impetus is at first only an 'indistinct multiplicity', a densely heterogeneous mass of interpenetrating tendencies defined by a forward thrust (TS 293; CE 258). Life evolves by radiating across a ramifying series of directions in response to the limitations that constrain the extension of each. Its essence is 'to develop in the form of a sheaf, creating, by its very growth, divergent directions among which its impetus is divided' (CE 99). Each new set of directions is formed like streams around an obstacle. They were each drawn from out of the original course, but they could not have been conceived through it in advance.

A ramifying system of rivulets retains the impulsion of the river that produced it, sacrificing only what is incompatible with the course of each new stream. Tendencies develop analogously as they bifurcate, suppressing what cannot be retained and extended across each new set of directions. Bergson formulates the idea as a law.

When a tendency splits up in the course of its development, each of the special tendencies which thus arise tries to preserve and develop everything in the primitive tendency that is not incompatible with the work for which it is specialised (CE 119).

The two laws of tendency development that appear in *The Two Sources*, and that I mentioned in Chapter 3, seem to have been generated via the bifurcation of this one law from *Creative Evolution*. Bergson conceives evolution as the unfolding of a single tendency according to this idea of specialisation. It is not unfolded linearly, across a unitary, progressively complexifying line of ascent. It is diffused across a widening web of trajectories, each of which represents the development of some aspect of the originating condition. These aspects or elements existed first in a state of interpenetration in a primitive tendency that preceded their dissociation across distinct lines of development. Divergent evolutionary trajectories may appear complementary to each other in certain respects, owing to the fact that the qualities that define them were qualities previously fused together and unified.

One example of this complementarity is the dependence relation between plants and animals. Generally speaking, plants derive their energy autotrophically (as do algae and many bacteria), by producing complex organic compounds through the photo- and chemosynthesis of simple elements in the sun, soil, and air. Animals cannot assimilate these elements unless they have been converted into organic compounds for them by autotrophs, or by other animals that owe them to autotrophs. Plants nourish animals, and animals are able to expend the stored energy they derive from plants in a greater number of directions.

If we imagine that early life forms faced the problem of how to obtain, suspend, and utilise free solar energy, it seems plausible that they tended towards both assimilating it directly, autotrophically, and expending it via locomotion as well, the way a heterotroph would. Such a form of life would not be capable of evolving far in the direction of locomotion, since it would be restricted by the necessity of manufacturing its own energy reserves. Evolution would have progressed by dividing the two operations across plants and animals, allowing each to elaborate one of them further. Thus 'while the animals evolved . . . toward a freer expenditure of discontinuous energy, the plant perfected rather its system of accumulation without moving' (CE 116).

The dependence relation that obtains between plants and animals should not be interpreted as the mark of a 'reciprocal adaptation', 'produced, in course of progress', in order to bring two initially discrete forms of life into a kind of useful alliance from the outside (CE 117). Complementary relations are better understood as having arisen 'from an original identity, from the fact that the evolutionary process, splaying out like a sheaf, sunders, in proportion to their simultaneous growth, terms which at first completed each other so well that they coalesced' (CE 117). Realised across divergent evolutionary lineages, these terms retain a virtual affinity to each other and so preserve 'an appearance of kinship' (CE 116). Complementarity is a manifestation in the actual of a past unity in the virtual (see Jankélévitch 2015: 123).

Bergson mentions *Euglena* as a possible example of a form of life early enough to pre-date the division between plant and animal tendencies (CE 116). Most species of *Euglena* have photosynthesising chloroplasts in their cell body. These chloroplasts enable them to derive energy through photosynthesis, like plants. *Euglena* can also surround organic compounds and consume them by phagocytosis, which is a heterotrophic operation. They represent a unification of the tendencies that Bergson attributed to plants and animals respectively. Though Bergson did not know it, *Euglena* owe their photosynthetic capacities to an endosymbiotic alliance with a green algae, which means that they unified the evolutionary strategies of plant and animal over time, through the reciprocal adaptation of two distinct organisms (see Gibbs 1978). This makes *Euglena* an unfortunate case study for Bergson's hypothesis, which is that 'the harmony between terms that are mutually complementary in certain points is not, in our opinion, produced, in course of progress, by a reciprocal adaptation; on the contrary, it is complete only at the start' (CE 117). To be fair, Bergson claims only that the *Euglena* 'may symbolize' the primordial tendency in which stationary autotrophy and locomotive heterotrophy were first fused (CE 116). That tendencies can be unified symbiotically should not surprise us if symbiosis is past unity made manifest.

Complementarity is an effect of the development of different sets of characters in two directions. The stable autotrophy of plant life complements the locomotive heterotrophy of the animal. If there was a tendency that enfolded both prior to their division, then one should find a primitive form of the tendency towards the locomotive expenditure of energy in the plant. It may be suppressed, or 'asleep', taken over by the development of the tendency towards autotrophic torpor, but preserved there nevertheless (CE 119). Adaptive benefit may contrive to awaken it. This is what accounts for phenomena of movement and heterotrophy in climbing and insectivorous plants, as well as for phenomena of fixation in parasitic animal species. What one lineage develops, the other suppresses, but does not eliminate.

Complementarity obtains when this inverse relation between development and suppression lines up in two or more species. Plants evolve the autotrophic capacities on which animals depend by suppressing the tendency towards locomotion and vice versa. The complementarity of different terms can be interpreted as a 'trace' or 'mark' of terms shared in common, though developed differentially (CE 118). This is what Bergson means by the claim that 'there is no manifestation of life which does not contain, in a rudimentary state – either latent or virtual – the essential characters of most other manifestations' (CE 106 tm). In any one of its forms, the characters of other forms can be found (CE 118). What is manifest are those characters 'whose activity does not obstruct the development of the elementary tendency' in each line (CE 119). What is virtual, left in a rudimentary state, are those characters that the elementary tendency had to suppress to extend itself further in each direction.

As tendencies bifurcate, they traverse a spectrum of particularisation. Each new set of directions is more specialised than the last, focusing themselves in narrower forms over time. At each stage, the development of a certain aspect of a tendency along one line is suppressed in the other. This spectrum originates in the most basic divisions in the history of evolution, such as those that define different kingdoms. Each tendency is then particularised further via the various strategies of each kingdom, which radiate through the differentiation of species. At this end of the spectrum, tendencies are specialised for the particular functions, body-plans, strategies, and behaviours distinctive of differently adapted populations. At their most particular, tendencies assume the form of the underlying generators or developmental pathways that account for the development of particular traits. It is not the case that the fully formed traits of one species are held dormant in another. What is held dormant is the state of the tendency towards the development of those traits prior to the distinction between the two species and the formation of the traits in one of them. It is the developmental pathway, or underlying genera- tor, that is dormant. The realised structure belongs only to those species that realise it.

When the same trait appears across divergent lineages, it is in one sense a result of parallel evolution, not convergence. What has been recovered is a dormant developmental pathway that was initially shared in common. In this sense, deep homology is the condition for the appear- ance of like structures on different lines. Adaptive benefit provides the opportunity for homologous pathways to materialise the same characters in each case. But it is not only the common derivation of a developmen- tal pathway that is operative in these instances. Bergson does not restrict himself to the claim that only species sharing a relatively recent common ancestor can manifest like characteristics by deriving homologous under- lying generators from it. He holds that all manifestations of life contain within themselves the characters of other manifestations in a latent state. The strength of Bergson's account for convergence is supposed to be 'proportional both to the divergency between the lines of evolution thus chosen and to the complexity of the similar structures found in them' (CE 55). This is a standard that parallelism and homology are hard-pressed to meet. It is a standard externalism invokes to motivate an explanation for convergence by adaptation to like environmental pres- sures. The likelihood that similar structures can be derived from a com- mon history declines according to their distance from a shared ancestor.

According to Bergson, what divergent evolutionary lineages have in common is older than the generation of any one of them, as the 'common impulsion' that is divided across each species is coextensive with evolution as a whole. The evolution of the underlying generators that account for the development of realised structures is nested within a history that precedes the distinction between kingdoms. Convergence is supposed to testify to this history in its entirety, as the still-unfolding event of life.

Convergent phenomena appear to arise out of homologous developmental pathways. The pathways, facilitated by adaptive benefit, appear to condition the development of convergent traits. Deep homology represents progress in the direction of a trended understanding of evolution. This is an improvement on the idea that selection pressure alone furnishes the traits from the outside, by acting on a pool of random variations. According to the internalist theory of homology, camera-like eyes do not arise out of a conjunction of random chance and differential mortality, but out of shared developmental potentials tending towards certain outcomes and materialising them wherever possible and beneficial. But to stop there, tracing realised structures back to the underlying generators they share in common and claiming that it was the generators that caused them, would be to terminate the developmental account too early. What of the constitution of the generators themselves? If they are selected for, then they must confer some adaptive benefit independently of the realised structures for which they are later enlisted. But if that were the case, then homology would represent an adaptationism merely relocated one temporal order earlier.

On Bergson's account, the tendency that manifests itself in the realised structure of a convergent organ is older than the developmental pathways that canalise it. The evolution of the eye is a movement, an event. Like all movements, it can be artificially arrested and spatialised in material form. In the adult organism, it is embodied in the structural complexity of the visual apparatus; in the developing embryo, it is formed via the differentiation of an initially equipotential domain; and in the evolution of the species, it is materialised through the developmental pathway that underwrites its structural eventuality. The material actuality of a developmental pathway consists in a genetic cascade that controls for the morphogenesis of the realised structure (Michaut et al. 2003). Pathways evolve through the intercalation of new genes via duplication and diversification on the one hand, and enhancer fusion on the other (Gehring and Seimiya 2010).

Together these processes are supposed to account for the evolution and complexification of the generator underlying the development of the eye in each of its instances. This generator is organised around *Pax-6*, the so-called 'master control gene' for the eye, which is expressed early in organogenesis and initiates eye morphogenesis. Other genes, such as lens genes, are recruited into the cascade over time to furnish the various generators that underwrite the myriad of complex eyes across the animal kingdom. The internalist claim that eyes are artifacts of parallel evolution refers to the homologous derivation of *Pax-6* and associated genes from a common ancestor (see Shubin et al. 2009; cf. Tomarev et al. 1997 and Yoshida et al. 2014). Deep homology indicates an internal directionality operative within processes of selection in that different lineages recruit the same basic sets of genes to manifest and improve upon the same structures.

Here is the Bergsonian position. The generators, or developmental pathways, are phases in the development of the functions that they help realise. Functions are embodied in organic structures, which are realisations of the pathways that underwrite them, and the pathways are realisations of the tendency towards the function. There is one movement, one tendency, oriented towards an immanent end – the function, vision. The genetic cascade that controls for the development of the eye is as much of a materialisation of this tendency as the finished structure of the visual apparatus is. Both represent the same event spatialised at different instants. The tendency towards vision is canalised through the developmental pathways before it is canalised through visual apparatuses, but it is older than and immanent to the constitution of both.

If homology is a correlate of shared generators, then Bergson's account of convergence is a non-homologous one. The parallel evolution of homoplastic traits from out of homologous generators is nested within the larger evolution of the generators from out of more elementary tendencies. When the convergent organ in question is the camera-like eye, the developmental pathway responsible for its parallel evolution is a genetic cascade controlled by *Pax-6*, but the cascade is a canalisation of the tendency towards vision that pre-dates its genetic constitution. Wherever that tendency is able to begin materialising itself, it will eventually come to be embodied in like developmental pathways, which realise like organic structures. Both the pathways and the structures are derivative on the tendency. It is the virtual lineage of the tendency that is common across all appearances of a like organ. That is what they share, at a level more fundamental than that of the developmental pathways enlisted in their generation.

c. Recollection and return

Bergson's explanation for how the same forms are canalised through different channels involves another dimension of his psychological interpretation, the memory model for tendency development (see Deleuze 2006: 77). This model consists of three ideas. The first is that evolutionary tendencies can be analogised with memory-images. Both are automatically preserved in duration, and both press into actuality from the past. The second is that the recollection of memory-images on the basis of present perceptions can be analogised with the materialisation of tendencies in response to adaptive circumstances. Just as like perceptions tend to attract like memory-images, like environmental problems tend to attract like solutions. The third is that the dynamic relationship between the past and its recollection in memory-images can be analogised with the dynamic relationship between virtual tendencies and biological traits. Just as recollecting the past reorganises it for future recollections, actualisations of evolutionary tendencies reconfigure the availability of those tendencies for future actualisations. Together these ideas complete Bergson's account of convergence.

1. All perception-images are retained as memory, subsisting virtually in the past. Habits inhibit memory, filtering what is not useful or relevant to the present. At the upper limit of recognition, where habits complete the objects of perception, there is no intervention of memory. Yet, 'our past psychical life is there' nonetheless, virtual. The retention of the past in the form of memory is a consequence of the indivisible unity of the duration of change. The past history of any trajectory of becoming is preserved along with it. We cut into the stream of time that is our duration, dividing away a past by rendering it dormant and focusing on present circumstances. The act of dividing away the past from the present virtualises the past. The virtualisation of the past is the constitution of some portion of lived time in the form of memory.

Memory is not an inert residue of the consolidation of the present. The past suffuses the present, completing it wherever and so far as possible. The present is what is left from the past when everything useless to some current situation is filtered away from it. According to the mereology of the virtual, the present is technically part of the past. Since 'something of the whole must abide in the parts', in every image the entire past is retained at some degree of contraction (CE 54). The present is the tip of the cone of the past, and every determinate memory is a contraction of the whole cone in the form of an image as well.

Tendency is like memory in this respect. Every determinate tendency is a part of the whole history of its development, and it retains that past as a virtual whole is instantiated in its actual parts. This is why autotrophy is supposed to subsist through the evolution of animal locomotion. The tendency realising itself across the differentiation of animal species is only divided away from the past it shares with plants from the perspective of some given animal. That animal selects away everything that is not useful to it, closing itself off from the evolutionary history that conditioned its emergence in order to focus its attention on its immediate environment.

Just as present perception succeeds only in virtualising the past without ever eliminating it, the constituted actuality of a given species succeeds only in virtualising the forms through which evolution previously passed. The perceptual present is nothing but the 'invisible progress of the past gnawing into the future', swelling as it advances (MM 194). 'Life', likewise, 'is a continually growing action', and each extant form is merely the crest of an advancing wave (CE 128).

2. The virtual existence of memory consists in its tendency to become present. Memory presses into the present. Even when it is inhibited by the continuous circuit of perceptions and actions, 'memory merely awaits the occurrence of a rift between the actual impression and its corresponding movement to slip in its images' (MM 113). It does that in most cases to help reinstitute the circuit between action and perception, feeding perception the images it needs to activate the right behavioural schemata. Memories are recalled either because of a resemblance they share with perception-images, or for their utility for some current system of actions. Memories are actualised in a series of stages 'by which this image [of the past] gradually obtains from the body useful actions or useful attitudes' (MM 168). While it looks as if the present is reaching backward into the past to find the image it needs to complete its prolongation into action, at the same time 'the virtual image evolves [from the past] towards the virtual sensation, and the virtual sensation towards real movement' (MM 168).

The virtuality of the past is a tendency to become present; it only needs the opportunity to do so. Like perceptions or present situations facilitate the actualisation of like memory-images in response. Habits and skills consist in the formation and regularisation of a circuit of some set of memory-images with some set of present perceptions. Recollection works similarly, as returning to one's childhood bedroom will tend to attract memory-images from the time spent there as a child. Memories do not need to be located in and drawn out from the past; they press into the present, awaiting an open channel. It is when no such channel exists that we find ourselves having to strain in order to recall.

Evolutionary tendencies are like memories in this respect too. Their actualisation is facilitated by adaptive benefit, which is analogous to present perception. As memories respond to present circumstance by inserting themselves as images, tendencies respond to material problems in the form of evolutionary solutions. Similarity in adaptive problems will tend to draw similar tendencies into actualisation. When the present situation is the problem of how to navigate a light-saturated environment, the corresponding 'memory' is the tendency towards vision. The adaptive pressure of light does not cause the evolution of the ability to see; it acts as a conduit for the actualisation of the tendency towards vision. That tendency is channelled through the solution to the problem, in the form of the eye. The structure of the eye is a material representation of the function of vision, and vision is the sensory facet of indetermination. Light is a navigatory problem, a problem posed to locomotion. Vision is its solution, not as an isolated capacity, but as an integral aspect of a sensorimotor system optimised for choice. The causal order of the evolutionary process is not linear but bi-modal, as adaptive problems are refracted through the virtual tendencies whose actualisation they facilitate and back into the organic solutions in which those tendencies are actualised. When Bergson claims that 'adaptation to environment is the necessary condition of evolution', this is what he means (CE 101).

Different perceptions can act as conduits for the same memories. A similar scent can bring the same memory-image into actuality no matter where it is encountered. One cluster of virtual memory-images is channelled through various perceptual circumstances. An analogous asymmetry obtains for the relationship between tendencies and their materialisation in different evolutionary situations. So long as those situations pose a similar adaptive problem, they can act as conduits for the actualisation of the same virtual tendencies (see Wong 2016: 74). What I am calling a 'conduit' or 'channel' for the actualisation of a virtual tendency, Bergson referred to as an adaptive 'occasion' for the 'release' or 'unwinding' of a virtual effect (CE 73). This is how to understand his remarks about the three forms of causality (see Riquier 2008: 301–5). Whereas adaptationism conceives selection pressure as the cause for the formation of a character by 'impelling' it, Bergson reconceives the causal efficacy of selection pressure in the form of an occasion for the actualisation of a tendency – by 'unwinding' or 'releasing' it – that manifests itself in a particular character. The difference is that while 'impelling' implies a direct relationship between cause and effect, 'unwinding' and 'releasing' imply only that the effect is given an opportunity to realise itself.

Externalism is partly right in that it is like selection pressures that explain the appearance of like organs across divergent evolutionary lineages. Selection pressures explain traits the way the topology of a hilly landscape explains the form of a road constructed through it (CE 102). The pressures act as contouring agents, providing necessary conditions and constraining what can be done with them. But they are not causally primary in the traits adapted to them. Internalism, too, is partly right in that it is like developmental potentials that explain the appearance of like organs on divergent lines. Those developmental potentials are formed under the pressure of adaptive circumstances without being explained by them.

Bergson's unification of these two positions consists in a metaphysical supplement. He supplies an account of how it is that developmental potentials arise in the first place in order to be contoured and canalised through selection pressure towards the generation of like traits in the second. The explanation is a modal one. Environmental problems act as channels for the actualisation of virtual tendencies. Actualised tendencies appear in the form of solutions, whose structural composition is conditioned by selection pressures. Their convergent similarity is not an effect of homologous underlying generators, but of the same virtual tendency canalised through different channels, like the same memory recalled under different conditions, conforming itself to them.

Convergent organs are actual solutions to similar environmental problems that channel the same virtual functions. The homoplasty of realised structures is a solution to similar adaptive problems, which is what externalism gets right. The convergent traits, far from being caused by those problems, are the developmental realisations of homologous generators, brought into actuality adaptively, which is what internalism gets right. Bergson adds that the homology of generators is a result of the canalisation of the same virtual tendency through different channels, evolving along different lines while retaining the same shared past in each. It is that shared past that can be recalled in new situations, like a memory-image, appearing in like traits when channelled for adaptive benefit.

Bergson claims that 'when we meet, on one line of evolution, a recollection, so to speak, of what is developed along other lines, we must conclude that we have before us dissociated elements of one and the same original tendency' (CE 118). He confirms his earlier wager that 'if the essential causes working along these diverse roads [i.e., evolutionary lineages] are of a psychological nature, they must keep something in common in spite of the divergence of their efforts, as school-fellows long separated keep the same memory of boyhood' (CE 54).

Shared histories are retained across divergent trajectories and can be recalled in each line whenever channels towards actualisation present themselves. Since it is the same virtual tendency that is being rechannelled in each circumstance, Bergson concludes that 'no matter how distant two animal species may be from each other, if the progress toward vision has gone equally far in both, there is the same visual organ in each case' (CE 96). This is because 'the form of the organ only expresses the degree to which the exercise of the function has been obtained' (CE 96 tm). The function is a virtual tendency; the organs through which it is canalised are actual forms, whose structures are conditioned by selection pressures. Like organs, channelling the same function, appear when selection pressures conspire to facilitate the actualisation of the same tendency in both cases. Those organs appear in comparably complex form when that tendency has been able to develop comparably far in both lineages.

No matter how complex, every visual organ is an apparatus for seeing. The tendency operative behind the convergent reappearance of like traits is not a tendency towards the formation of those traits themselves, but a tendency towards the realisation of the one function canalised through each of them. 'Parallel evolution shows', as one scholar says, 'that organisms may be predisposed to evolutionary tracks that are definable more in terms of function than of structures' (Wesson 1991: 191). The complexity of each form is a correlate of the progress made by the tendency towards its realisation along those 'tracks'. As that tendency advances, it is canalised through increasingly and analogously complex characters. 'Species may develop toward similar purposes in physiologically and organically different ways' (Wesson 1991: 191). If eyes of like complexity appear in molluscs and chordata alike, it is because the same tendency towards vision has made a similar progress in each phylum.

One way to make sense of this progress is to ask what the species that realise it have in common, other than their eyes. Besides vertebrates and cephalopods, such as the octopus, camera eyes are convergent in Cubozoan box jellyfish, *Portia* and *Dinopis* spiders, and some Gastropods, including snails of the *Strombus* genus as well (see Morris 2003: 151–7). These species are agile, coordinated, and fast. The spiders are dexterous hunters with impressive motor control. The box jellyfish are skilled swimmers, manifesting complex, visually guided response behaviour (Skogh et al. 2006). The snails are adept at registering threats quickly and, somewhat surprisingly, rapidly evading their predators as a result. When Bergson says that similar visual organs represent similar levels of progress in the realisation of vision, he means that each species has evolved to a similar level of coordination between its ability to perceive and its ability to act with precision in response. Vision is not sensory, but sensorimotor. A sophisticated ability to see is incomprehensible apart from a complex capacity to act on visual data. Sophisticated vision and sophisticated motor control are two facets of the same fact. It is towards that fact, the intensification of action, that evolution is oriented. The tendency channelled through the visual apparatus is a particularised variant of the tendency towards action, indetermination, or choice.

Bergson mentions the convergent evolution of complex vision in humans and octopuses. It is among the most impressive cases of convergence. Our last common ancestor was a sightless worm in the Ediacaran era, 500 million years ago. Since then we evolved eyes comparable in structural complexity and functional sophistication. Octopuses are visually guided predators with close, responsive control of their tentacles. Their camera vision facilitates their dexterity and is indissociable from it. Perception is virtual action, no matter the species. Sophisticated perception is a correlate of action capacity. Externalists are wrong to think that whether camera vision is adaptive for a species depends on selection pressure or environment alone. It also depends on the neurology and morphology that coordinates complex vision with acute motor control, coupling perception with action. When the conditions are right, when the channel towards vision is available and beneficial, complex eyes can become increasingly inevitable, no matter how divergent the trajectory or dissimilar the environment.

3. There is one final facet to Bergson's psychological model for convergence. In Chapter 2, I called it 'dynamics'. It consists in the bidirectional relationship between the virtual and its actualisations. The actual is not only a product of the virtual, but a means by which the virtual is reconfigured. The relation between memory and its recollection in the form of memory-images runs from the past towards its contraction in actualised images, and from memory-images back up towards the whole contracted in them. Memory-images generated in the present and added to the past reshape the whole because the whole is contracted in each of its parts. Every part implicates the whole, which means that the addition of new parts to the whole has to reconfigure it in order to contract it anew. The more often a cluster of images is recalled, the more readily and regularly the same memories tend to be recalled again. The layout of the virtual landscape is recontoured to enhance the actualisation channels available to select images, dynamically conditioning their actualisation (see Gunter 2007: 39).

Evolutionary tendencies are like memories in this last respect. They exist in a state of interpenetration prior to their dissociation across history. In that state, they are internally related to each other. Internal relations are relations that constitute their terms. Changes in internal relations imply changes in each related term. External relations, on the contrary, are interchangeable, non-constitutive relations that obtain between independently individuated actualities (see Whitehead 1967: 135; cf. Bradley 1893: 142, 364, 392). Changes in one term of such a relation do not change the others. Neither does a change in the relation itself impact on the related terms. Virtual tendencies are internally related to each other because they are not already individuated before being dissociated. It is the act of dissociation, actualisation, that individuates them (CE 257). The act of actualisation also rebounds on the whole as an internally related nexus of other tendencies. As Pete Gunter observes, if life is a virtual whole, then 'expressions of one aspect of the impetus would have impacts on the others' (2007: 39; cf. CE 250). Life is reconfigured through each of its actualisations, evolving as a whole in tandem with the evolution of each of its dissociated parts.

The consequence is that each expression of evolutionary tendency in the form of a bifurcated set of tendencies creates actualities without having prefigured them and recreates the virtual conditions for future actualisations as well. As Gunter observes, though 'Bergson gives us good reasons to hold that the possible does not precede the real [i.e., actual]', there are nonetheless on his terms 'multiple dynamic factors that do' (2007: 39). As certain memory-images become increasingly available for regular recollection over time, the actualisation of certain tendencies, or certain elements of them, become increasingly probable wherever minimal conditions obtain. Every actualisation of a tendency in one form increases the probability that it will be actualised in a like form again. This probability increase is an index of the dynamic differentiation of the past in the form of what Deleuze calls '*regions*, *strata*, and *sheets*: each region with its own characteristics, its "tones," its "aspects," its "shining points" and its "dominant" themes' (1989: 99). Convergent organs are like the actual correlates of the 'shining points' of the virtual past. By tending towards the same points and themes more often than to others, evolution regularises itself over time, even as it ramifies. The paths followed by each line converge upon the same small set of traits such as the eye wherever feasible, like an organic refrain.

Regularisation, or refrain, is a correlate of evolution's directionality. Since organisms retain the history of life within themselves as a single event, there is no limit to where and how the same forms or functions will continue to reappear. The evolution of convergent traits is limited only by its own unfolding history, which tends to regularise itself over time, in much the same way that a wide variety of possible behaviours tend to order themselves into a routine, then a habit, then a set of reflexes as they are reiterated again and again. Far from being an explanatory challenge or exception to an otherwise chaotic procession, convergence reveals the unity and directionality of the evolutionary movement. From this perspective, widespread convergence is not only comprehensible but unsurprising and perhaps even inevitable.

Bergson describes 'a tendency to associate' that is opposed, complemented, and completed by the 'tendency to individualize', from unicellular organisms to social formations (CE 259–61). The tendency towards individuality is the tendency towards the determination of relatively unified organisms in contradistinction to each other and their environments. It is a materialising tendency. Yet, 'among the dissociated individuals, one life goes on moving . . . as if the manifold unity of life, drawn in the direction of multiplicity [i.e., of individuals], made so much the more effort to withdraw itself into itself' (CE 258–9). The tendency towards association describes this refrain-like effort of life to 'withdraw itself into itself', manifesting unity against the exigencies of organic individuation. This tendency predominates in the formation of 'microbial colonies' and symbiotic alliances through the association of individuals in social arrangements all the way up the scale of complex life (CE 259). It is predominated over by the tendency towards individuality wherever associations, once coordinated, are 'melted' into new individual forms – such as in the integration of endosymbionts in the generation of more complex organisms and in the constitution of a kind of superorganism from out of a society of associated individuals, such as in a colony of insects (see Sagan 1967 and O'Donnell et al. 2015). Every new individual is material for a new association in turn, for 'the evolution of life in the double direction of individuality and association . . . is due to the very nature of life' (CE 261).

Life strives to return to itself wherever it can, manifesting as much of its virtual unity as possible, but the route backwards becomes increasingly circuitous as selection pressures divert an initially 'immense wave' into the winding waterways of ever finer 'rivulets' (CE 250, 54 tm; cf. Freud 1961: 46). Unity is a past that can never be made fully present, as life is caught up from the beginning in processes of differentiation. This may be what Jankélévitch means by the enigmatic contention that life is 'not a promise but rather a reminiscence' (2015: 123). The whole is forever behind the ongoing evolutionary advance of its parts. Tendencies diverge. Forms proliferate. Wherever possible, they also tend to associate, unify, and resist the tendency towards dissociation, individuation, and independent development. Convergent characters might be artifacts of the same tendency towards unity predominating over the material demand for division, manifested not in the association of already-extant individuals, but in the re-actualisation of a 'common element' across otherwise divergent trajectories (CE 54). By retaining 'something in common in spite of the divergence of their effects', tendencies that canalise for convergent characters also reiterate the unity of life against its dissociation into parts (CE 54).

Why, then, does the same tendency become canalised in so many different circumstances, manifested across similar orders of biological traits? Partly because of the adaptive potential those traits have, given a set of selection pressures common to each circumstance. But also because the tendency towards vision has become an organic refrain. The virtual has come to be calibrated for vision to insinuate itself in a visual apparatus wherever minimal conditions allow. It is 'recalled' by adaptive circumstance like a memory, becoming one of the 'shining points' of the virtual past. Vision is a tendency, just like memory is a tendency – not only to become present, but to realise itself as far as present circumstances permit. It manifests itself with a likelihood that increases according to its own history of past manifestations. To borrow a term from chaos theory, vision begins to act as an 'attractor' for the evolution of self-movement over time (see Wesson 1991: 144 and Wilson et al. 2007: 889). It draws developmental trajectories towards it, stabilising and trending an otherwise random proliferation of adaptive traits around the same order of forms.

What I have been calling the 'shining points' of the virtual are reminiscent of Daniel Dennett's 'good tricks'. Dennett re-describes common evolutionary solutions, such as camera vision and animal metabolism, as 'good tricks' for solving ubiquitous problems. They are common because they are efficient and successful. Dennett likens 'good tricks' to 'beacons in Design Space, discovered again and again, by the ultimately algorithmic search processes of natural selection' (1995: 144). There is a sense in which the possibility of a 'trick' like the camera-like eye precedes its actualisation in particular organisms. Evolution converges on the eye not only because the eye is a possible solution to a common set of problems, but because it is a particularly successful one, and the 'algorithmic search processes of natural selection', set to find the most optimal and efficient solution in the Design Space of all possible solutions, are drawn towards the eye again and again as a result.

It might be objected that by calling the tendency towards vision one of the 'shining points' of the virtual past, I am reproducing the same metaphysics of possibility that I criticised Dennett for endorsing. What is the difference between a 'good trick' acting as a 'beacon' in 'Design Space' and a 'shining point' acting as an 'attractor' in the virtual? According to the Bergsonian account, the eye as an organic form is not a possibility before it is evolved.[3] Every organic instance of a visual apparatus is a contingent event. The eye is not a 'good trick' for resolving a common set of environmental problems. There is no 'eye' as such, neither possibly nor actually. There are only the variously complex material manifestations of a virtual tendency realising itself in the direction of increasing capacities for vision, embedded in different sensorimotor systems and body plans, such as those of jumping spiders and octopuses. The realisation of this tendency consists in the sensory presentation of indetermination. The development of the tendency towards indetermination 'takes directions without aiming at ends' (CE 102). It appears in the form of perceptual choice to locomoting animals. Visual organs seem to converge only in species that have to resolve information at high speeds for locomotive purposes (see Zylinksi and Johnsen 2014). Wherever those circumstances obtain, the tendency towards indetermination opens a channel for the actualisation of vision as an intensification of the ability to move. Its canalisation is contingent on the physiology of the species that manifest it and the adaptive potential it presents given the pressures under which it is evolved.

Natural selection does not discover the eye as a beacon in Design Space, actualising it over others because it is the best possible solution to a wide set of problems. What 'shines' is not a possible form, but a functional tendency. The tendency becomes increasingly available for actualisation as it is regularly actualised over evolutionary time, though the forms that manifest it are unforeseeable. They converge, because they each represent the progress made in the development of one tendency and so appear in structurally analogous form wherever that tendency advances to a similar point. That, too, is unpredictable.

d. Conservation and constraint

Where does the science stand on the issue? Is convergence an accident, the chance result of similar pressures acting on similar pools of variations, or is it an indication of unity and direction? Is it true that Bergson's 'claim that cases of convergent evolution are to be explained in terms of an initial impulsion of life that has persisted across divergent lines can only remain highly speculative', better regarded as a regulative idea than a viable postulate, as Ansell-Pearson has suggested (2002: 92; cf. Ricqlès 2008)? I conclude by introducing and explaining two ideas from the contemporary study of convergence that seem to have ratified at least some aspects of Bergson's account. These ideas are conservation and constraint.

Conservation refers to the retention of similar or identical sequences in either DNA and RNA or proteins across species.[4] Sequences can be more or less highly conserved depending on how long they have remained relatively uniform. One common example is the homeobox genes, which induce and direct the formation of body structure in early embryogenesis, dictating where arms, legs, or eyes will form in young embryos. Homeobox genes are found across animals, plants, fungi, and some single-cell eukaryotes. This suggests that they are conserved from their molecular origin in an ancestral nucleated life form, having been recruited to play an array of morphogenetic roles in an array of eukaryotic lineages since. Base sequences tend to be conserved when they prove capable of regulating a variety of developmental processes. Their utility for selection is decoupled from their implication in the production of any specific trait or function and telescoped outward into the surrounding traits or functions in whose production they can play a formative role. Sequences tend to be conserved to the extent that they can be recruited in the formation of a large set of body plans.

At a sufficient scope of regulatory utility, conserved sequences become part of what theorists call the 'genetic toolkit' for widespread development. The component sequences of the toolkit have been discovered in animals as diverse as worms, flies, and human beings. Jacob famously claimed that most of morphological evolution occurs through the 'tinkering' or 'bricolage' of the components of this toolkit, rather than the evolution of new components *de novo* (1997: 1164). This is purportedly why the Cambrian explosion of new forms post-dated the origins of the genetic toolkit. Evolutionary novelty is not strictly dependent upon chance variations, since new forms and functions can be generated through the rearrangement, or tinkering, of a small number of sequences. Evolution tends to elaborate new uses for conserved processes instead of inventing the building blocks for each new construction in the formation of the new construction itself. It should be no surprise to find similar genetic sequences and developmental processes at work in shaping the divergent morphologies of entirely different kingdoms.

When convergent characters have evolved on the basis of the same toolkit, and when that toolkit has an ancient origin, it represents what theorists call a 'deep homology' underlying the ostensibly independent lineages. When two lineages derive the same trait from a common ancestor that had it, those traits are homologous to each other. Conservation allows two lineages to derive the same trait from a common ancestor that did not have it. What each lineage retains is a developmental process from the last ancestor common to both. That process is eventually recruited in the production of the same trait in both lineages, which means that the trait is technically homoplastic, even though the genes responsible for its formation were not themselves independently evolved. This combination of trait-homoplasty with sequence-conservation testifies to a deep homology beneath the non-homologous convergence of the traits themselves. Conservation provides a form of historical unity retained beneath divergences in lineages operating upon the basis of conserved sequences. In some instances, the conservation can be so old, and the homology so deep, as to reach all the way back to the last universal common ancestor of all life.

I mentioned that the explanation of convergent characters by deep homology actually effaces their convergent status, recoding them as parallel. Bergson's account is different from deep homology for this reason, but there is a profound affinity between the two. First, and negatively, if the appearance of like characters across diverse lineages is a result of deep homology, then Bergson's critique of the adaptationist explanation for convergence appears to have received some scientific legitimation. Second, and positively, Bergson shares with deep homology an appreciation for the efficacy of the unified history of life. Third, both positions accord a form of directionality to evolution, and find that convergence exemplifies it.

Deep homology theorists agree that the convergent appearance of like characters is not a direct result of the action of like selection pressures and nothing besides. Convergence is not an accident, the chance occurrence of independent lineages falling into similar adaptive solutions in response to overlapping environmental problems. If adaptation is the 'cause' that explains convergence, it cannot be an efficient one. Adaptive pressure is better understood as performing the role of what Bergson refers to as an 'occasion' for the 'release' or 'unwinding' of the convergent quality instead (CE 73). Convergent characters are not passive reactions; they are the overlapping elaborations of a shared set of conserved pathways. Adaptive pressures provide the 'occasion' for those pathways to be realised in the 'unwinding', or developmental formation, of the same characters without causing them directly. This is what deep homology is supposed to demonstrate, by revealing beneath the convergent evolution of the eye the same set of regulatory genes conserved in the elaboration of the organ in each lineage. If this is true, then Bergson's critique of adaptationism seems to have received the empirical endorsement that it lacked when he initially articulated it.

There is an affinity between the explanation of convergence by deep homology and Bergson's own positive alternative to adaptationism. To be sure, deep homology does not invoke any explicitly formulated variety of 'shared impulsion', but its vocabulary of common developmental potentials, conserved and recruited for the formation of diverse morphological plans, appears close. Both positions discover an underlying unity beneath the appearance of non-homologous, independently evolving trajectories. Deep homology understands this unity in the form of developmental sequence and genetic network. It posits only a small set of sequences and networks, and finds that they have been conserved from their origin through the proliferation of species by being rearranged and repurposed in novel forms. These are what generate the plurality of diverse characters. The unity is not left behind by the plurality that supersedes it; it insists through and conditions the emergence of each new form. The convergence of like characters in unlike lineages reveals this deep unity. It does so even across divisions in kingdom, a key contention of both *Creative Evolution* and evolution by deep homology alike.

Bergson thought that analogies obtaining between *unlike* characters across different kingdoms might testify to the unity of evolution as well (CE 59–60, 87). On his hypothesis, 'the same impetus that has led the animal to give itself nerves and nerve centres must have ended, in the plant, in the chlorophyllian function' (CE 114). It is not only convergence that Bergson's account is supposed to explain, but the analogies between divergent characters as well, since even in divergence, the one shared impulsion of life is supposed to be manifest. There are two fronts to the contemporary study of this suggestion: the conservation of the same 'toolkit' across plants and animals, and the 'neurobiology' of the plant chemical communication system, which treats it on an analogy with the animal nervous system (see Heil and Karban 2009, Heidel et al. 2010, Karban 2015, and Karban et al. 2000). The idea is that living things are unified, in all their manifest diversity, not only historically, with reference to a shared origin, but internally and developmentally through each of evolution's divergences as well.

Last, deep homology shares with the Bergsonian account the idea of trend or directionality. Convergence is an in-built feature of the development of new forms for both accounts. They agree that evolutionary novelty is not reliant upon a randomly generated pool of variations filtered by selection for their utility. Evolution is trended by the developmental potentials that it conserves from its history and patterned by their realisation. Not all variation is chance variation, and not all realised structure is the direct result of selection for adaptive benefit. At least some variation is the result of conserved sequences being tinkered with and enlisted in new processes, which imparts a form of evolving directionality to evolution as past potentials shape future trajectories.

For Bergson, the source of trend is a tendency towards indetermination. For deep homology, it is the patterns and networks of genetic sequences acting as developmental potentials for the formation of different sets of body plans. Both hold that evolution is at least partially and powerfully non-random. Since evolution derives directionality from the conserved history that its lineages share in common, convergence is likely to occur. Evolution is trended by the common potentials that divergent lineages conserve, which means adaptive potential need only align for two species in order to make the realisation of those potentials in the form of like structures increasingly probable.

Conservation, and the associated idea of deep homology, provides one example of the attestation of contemporary science to a number of Bergsonian tenets. Adaptation is not a positive agent, but a contouring force in the unfolding of evolution. It does not act upon the chance proliferation of variations, but on the trends imparted to evolution by the developmental history that it retains. There is a deep unity underlying the plurality of life forms, revealing itself in the convergent evolution of like characters across divergent lineages.

The idea of developmental constraint provides a second example of how the recent study of convergence might support some aspects of the Bergsonian account.[5] The explanation for the recruitment of conserved sequences in the deep-homologous development of homoplastic traits is usually furnished in large part by way of the idea of developmental constraint. Constraints are typically defined as 'biases on the production of variant phenotypes or limitations on phenotypic variability caused by the structure, character, composition, or dynamics of the developmental system' in question (Maynard Smith et al. 1985: 265). If one imagines a 'morphological space' consisting of all possible phenotypes, a large range of them will be impossible to develop for any actually existing species, not because they are adaptively harmful, neutral, or otherwise suboptimal, but because there are material factors that limit which characters can be manifested. Within the narrower space of what is physically feasible, there is a further range of forms that, while not strictly impossible, are rendered highly unlikely, no matter their adaptive utility. The non-adaptive factors that limit the morphological space down in terms first of possibility, and second of probability, are called constraints. They include all the cellular, developmental, and morphological properties that work to restrict viable evolutionary trajectories.

Constraints can serve both negative and positive explanatory functions. Negatively, they curtail the relevance of adaptive utility in the explanation of observed phenotypes. Selection can only explain why some forms are inherited over others for a narrow window of viable variants. By limiting the pool of possible forms down to those that can be realised, constraints account for the distribution of variations upon which selection can act. Adaptive benefit explains the rest, but it can only do so once the relevant constraints have been enumerated. This negative function is clearest in the optimality model for adaptation. This model allows researchers to make predictions about the optimal characters of some phenotype, given a set of selection pressures and environmental challenges. If selection were the only force at work in accounting for phenotypes, then every character should be optimal for the problems it solves. But rarely do the characters first predicted match the traits observed (see Wesson 1991: 88–96). This is because there are constraints on which states can be realised. Considered in abstraction from these constraints, selection regularly appears to result in suboptimal traits. The researcher has to build the constraints into the model.

According to its negative function, constraint is constraint on selection, limiting the characters available for selection to select.

Constraints function positively too, by acting on the production of heritable variation in development. Constraints open specific sets of available developmental trajectories. The realisation of those trajectories in certain forms does not indicate anything about the fitness of those forms for the environments in which they will be selected for or against. Constraints can also result in 'spandrels', as Gould and Lewontin famously argued. A spandrel is an example of the adaptive evolution of one character resulting in another non-adaptive character as a developmental by-product. In such cases, the development of the non-adaptive character is positively constrained (see Pearce 2011).

The dual nature of constraint parallels that of canalisation. Limitation in the developmental domain is a generative operation, more like facilitation than restriction. It is an enabling condition, not only a check on possibility. 'Developmental theories', in one theorist's words, 'are generative theories' (Amundson 1994: 570).

One term for the roles played by constraint in development is 'bias'. Words like 'trend' and 'pattern' are also regularly employed in the literature. The idea is that embryological processes reliably produce forms within a certain range of variation (see Amundson 1994: 565–9). The production of form is 'biased' by developmental constraint in certain directions, resulting in an array of predictable outcomes (see Zalts and Yanai 2017). The pathways for the development of these structures are what constraints act upon. Structures are correlated with a variable number of developmental possibilities. Selection selects only for particular traits, not the new directions whose possibility they facilitate, but in so doing it retains those possibilities as a by-product. While both feathers and hair may have been initially selected for to aid in the regulation of body temperature, they involve different developmental possibilities. Only feathers open a pathway towards flight. Flight is an evolutionary novelty, a new quality, enabled by the morphology of feathers, but not selected for in advance. By opening pathways for the development and realisation of new potentials, constraints impart a non- or pre-adaptive trend to the production of viable variants.

Convergence is a result of the elaboration of new forms on different lineages in parallel directions. When Gould set out to reconceive the convergent evolution of the complex eye as an instance of parallelism, he invoked both 'the pre-existence' of 'developmental pathways' as well as the 'positive', 'internal constraint[s]' that act in order to 'strongly facilitate the resulting, nearly identical' visual apparatus in each instance (2002: 1128–9).

It is the conserved sequences, by presenting a definite set of developmental pathways, that act as positive constraints on the evolution of the convergent trait. For two species to conserve the same sequences is for them to share the same channels towards the same characters.

When positive constraints result from conserved sequences, it seems to follow that evolution is patterned by long-term tendencies. Long-term tendencies are the future-oriented analogues of sequences conserved from the ancestral past. Their directional realisation can be self-reinforcing as species evolve in certain ways. Once a photoreceptive collection of cells recruits the *Pax* genes in order to complexify further, the process picks up a kind of momentum, tending with more focused direction towards an increasingly complex eye. We should not have to consider every improvement the improbable acquisition of a chance mutation that builds towards the more complex structure because of its utility alone. Constraints imply both direction and pattern, as given structures specify for future trajectories as well as for associated traits, which allows them to 'vary in an integrated and functional manner', as one theorist puts it, and in a definite direction over time (Brigandt 2015: 313). Organisms do not consist of aggregates of independently selected chance variations, but of interacting developmental pathways whose trajectories are laid out ahead of them on the basis of the potentials that they conserve.

Orthogenesis is an immanent effect of the insistence of the past in the present of a process of change. It is neither reducible to the action of selection, nor does it require the postulation of an external principle. The trends that regularise the evolutionary process are not derived from pre-existent ends, but from what Bergson called a '*vis a tergo*', a force from behind, developing through divergent channels as it is differentiated over time (CE 103). 'One assumes', says Robert Wesson, 'that the octopus eye resembles the vertebrate eye because there are not many ways to make a cameralike apparatus' (1991: 189). If this is true, it is because there are not so many channels for the development of the complex eye. If that is true, it is because those channels are outlined in the form of developmental potentialities that constrain how and in what forms they can be realised. If that is true, it is because evolution conserves itself over time, re-employing the same sequences in an increasing variety of developmental processes. This demonstrates, finally, that complex eyes are not the result of adaptation building organs in the only way that they can be built, but rather that the history of life canalises its own evolutionary unfolding, deriving a set of directions as potentials from the past and realising them along a select number of channels that lead to a similar set of organs.

'It is not that evolution proceeds toward a goal', Wesson writes, 'but that it carries on in a direction that has been adaptive in the past' (1991: 192). 'This implies', as a result, 'something like orthogenesis' (1991: 194). Orthogenicity is derived from the past, from a shared history inherited in the form of conserved developmental tendencies that facilitate a definite set of available trajectories for the evolutionary future. The result is the appearance of trend, pattern, and coherence; not chance, accident, and mechanical reaction. The appearance is confirmed by 'the development in different groups of similar improbable organs and instincts' (Wesson 1991: 192). That is Bergson's position more or less exactly.

Conclusion

Life evolves from the virtual through the material channels of phylogenesis. It is an event, trended by virtual tendencies materialising themselves across history. Materialisation occurs through the medium of individual organisms, each of which instances a composition of temporalising and spatialising tendencies. Their evolution over time is testament to the predominance of the temporalising tendency on a higher scale. Bergson accords it a 'psychological interpretation'. That interpretation, rendered as virtual tendency, should be situated within the context of orthogenesis. Evolution derives its directionality from the history each of its diverging lines shares in common. That history is dense with tendency. The realisation of those tendencies is constitutively external to all individuals, subsequently internalised in the development of particular lineages, and then inherited and expressed in parallel through homoplastic traits. The connection between Bergson's account of convergence and his reformulated finalism resides in canalisation, a concept I draw from his two images for development, the canal and the hand thrust through filings. The images depict a relation between tendency and its material manifestation. Those manifestations are like parts standing in a relation of dissociation to the virtual unity of life as a single event, an open whole.

Here are the implications for the contemporary study of evolution. Adaptation is not the sole or only agent at work in evolution. In fact, it is better understood as a negative filter or contouring force operating on a more primary production. Selection does not act only upon the chance proliferation of variations, but rather on the directionality imparted to evolution by the tendencies that trend it and the developmental history that it retains (through conservation and developmental constraint). Finally, there is a deep unity underlying the extant plurality of life forms, which reveals itself in the convergent evolution of like characters across divergent evolutionary lineages. Each of these claims remain challenging, if not viable, for the contemporary study of variation, homology, and convergence. Bergson's 'psychological interpretation' of tendencies, and their virtual modal–mereological position, should not distract from that fact.

Concluding Remarks and Future Directions

There are a few general ideas that have informed the shape of this book. The first is that Bergson's philosophy can be reconstructed as a work of metaphysics in something like a unified, systematic fashion. Deleuze may have popularised this possibility, employing Bergson as an ally against the idea that metaphysics had come to an end and had to be overcome as a result. In an interview, Deleuze identified himself with a metaphysical approach to Bergson's philosophy. 'Bergson says that modern science hasn't found its metaphysics, the metaphysics it would need. It is this metaphysics that interests me' (qtd. in Villani 1999: 130). It is this metaphysics that interests me too. The idea of articulating a metaphysics adequate to current research has oriented my understanding of Bergson's project both in its own right and in its ongoing relevance for contemporary thought.

The second idea that informed the book is that Bergson's metaphysics receives its fullest extension and deepest instantiation in a philosophy of evolution, which that metaphysics underwrites and by which it is regulated. I work this idea out across each chapter, but it is clearest in the last half of the book, especially in the account of the *élan vital* as a psychological image for a metaphysical conception of tendency. I think that Bergson's philosophical output should be read outward from *Creative Evolution*, such that *Time and Free Will*, *Laughter*, and *Matter and Memory* represent incursions into a particular high-dimensional recapitulation of a more basic logic of tendency, multiplicity, and time operative first on an evolutionary register. *The Two Sources of Morality and Religion* represents the regulative conversion of a movement of speciation into a programme for revolutionary sociocultural and technological advance through the medium of individuals and the creative emotions that they are capable of channelling. Bergson's mystic, on this account, would stand towards the species the way the evolutionary movement stands towards its artificial stopping points: as a tensional thrust onward, fracturing consolidated forms onto the open future of creative experimentation. Bergson's infamous and infamously difficult theory of intuition – though I have left it to one side in this book – might be reconceived accordingly as well. I will say a few words about that below.

The third idea is perhaps the most obvious. Bergson should be studied not only for his historical importance, but for his contemporary relevance as well. Though his work has been out of fashion for many decades, it is animated by problems, informed by critical tools, and structured through insights that remain pressing for the philosophy, the biology, and the philosophy of biology of today. This is by no means a unique view. Many commentators share it. In this book I have tried to provide it with more positive biological content, showing exactly how and in what way particular aspects of Bergson's thought might prove relevant for particular theories and problems in the life sciences, and exactly what Bergson's thinking would entail in the form of a scientifically literate philosophy of biology.

★ ★ ★

I would like to make a final Bergsonian gesture by locating within this conclusion a set of open directions for future development. I will focus on four. The first is a concept of symbiosis that remains unexcavated within *Creative Evolution*. The second is a reinterpretation of the theory of intuition on the basis of symbiosis. The third is an elaboration of Bergson's philosophy of technology. The fourth, finally, is the relationship between Deleuze and Guattari's symbiosis-based evolutionary philosophy, with its focus on the becoming-together of wasp and orchid, and Bergson's discussion of the 'sympathy' operative between the wasp and the caterpillar in instinctive rapport.

1. It goes more or less without saying that evolution today cannot be thought adequately without deference to the near ubiquity of symbiosis in life on earth. From phylogenesis to immunological integrity, from the origin of species to the ongoing individuation of organisms, symbiosis is coming to be understood as an increasingly indispensable facet of evolutionary biology. It tends to be well-accepted that Bergson lacked a theory of symbiosis, and this lack is sometimes taken to constitute an objection to the contemporary relevance of his account. I think that is wrong, and that there is a subterranean line of symbiotic thought running through *Creative Evolution* and organising it from within.

My position that there is a theory of symbiosis to be found in *Creative Evolution* involves the adoption of another perspective on the virtual. The insistence of the virtual unity of life through its particularised parts is revealed not only in cases of convergence, but in cases of association and alliance as well. Symbiotic rapports are best understood for Bergson not as the external accord between two distinct individuals, but as the manifestation of common virtual tendencies through them. This account can be derived from Bergson's discussion of the biological 'sympathy' evidenced in the parasitic relationship between solitary wasps and the caterpillars that they seem to know instinctually how to paralyse. My wager is that sympathetic relations – non-epistemic relations of access – reach back up the actualising, dissociative evolutionary process through which tendencies that originally coexist are differentiated and then determined in separate forms. By reaching back up into this field of coexistences, the participants of a sympathetic relation coincide at a virtual point prior to the event of their dissociation. Sympathetic rapport is like evolution introjected and replayed in reverse: from difference back to unity, from the actual to the virtual and back. My second wager is that sympathy – as a fold back from the actual domain of dissociated tendencies towards the virtual field of coexistences – is the means by which Bergson's philosophy of evolution can be made to account for symbiotic relations. What is ultimately at stake in Bergson's discussion of instinctual rapports is a theory of being- or becoming-together, that is, of symbiosis. Bergson calls it 'sympathy', but it is the same idea nonetheless.

2. Bergson's theory of intuition can be reconceived on the basis of the nascent conception of symbiosis. The way into this connection is provided by *Creative Evolution*'s definition of intuition as 'instinct that has become disinterested, self-conscious, capable of reflecting upon its object and of enlarging it indefinitely' (CE 176). There is an evolutionary relationship between instinct and intuition. The latter appears to be a consequent eventuality of the development of the former away from its original interestedness in the specific object of its application. Intuition seems to be achieved via the de-particularisation of instinctual access. After suggesting that instinctual rapports penetrate the interior of life in a manner that is constitutively foreclosed to the capacities of intellection, Bergson claims that it is to that 'very inwardness of life' that intuition is supposed to lead us (CE 151, 165, 176). The suggestion appears to be that while both instinct and intuition are modes of access to the interiority (or virtuality) of the evolutionary movement, instinct is tied to a specific object while intuition is capable of an indefinite and reflective extension.

In the 'Introduction to Metaphysics', Bergson defines intuition as 'the sympathy by which one is transported into the interior of an object in order to coincide with what there is unique and consequently inexpressible in it' (CM 135). The promise of intuition is that it will provide the means by which to circumvent spatialisation and thereby allow for an entrance into some object's temporal or virtual reality. If sympathy is a mode of transindividuation, a fold of distinct actualities back onto their co-implicated virtualities, then intuition is intellectual sympathy, or a sympathetic relation enacted on purpose by a knowing subject.

The obstacle to intuition, its singular difficulty, is intrinsic to the evolved nature of thought itself, which tends towards intellection as its natural end. Intuition requires the frustration of this adaptive tendency. Bergson holds that the mind is capable of more than what it was evolved for:

> it can be installed in the mobile reality, adopt its ceaselessly chang-ing direction, in short, grasp it intuitively. But to do that, it must do itself violence, reverse the direction of the operation by which it ordinarily thinks, continually upsetting its categories, or rather, recasting them. (CM 160)

As a consequence, '*to philosophize* [that is, through intuition] *means to reverse the normal direction of the workings of thought*' (CM 160). Intuition as thought without intellection is closer to instinct than it is to typi-cal human knowledge. Like instinct, intuition relates to its object not as a constituted actuality but through the virtual past contracted in it. Intuition is a bid in favour of time, which involves a resistance to the tendency to spatialise. Just how this is to be achieved, and just what it is supposed to look like once it has been, is a still-open question.

3. The idea that there is a philosophy of technology to be elaborated on the basis of Bergson's account of the instinct-intelligence division belongs originally to Canguilhem, who called *Creative Evolution* 'a treatise of gen-eral organology' (2008: 174, n. 64). Canguilhem had in mind the etymo-logical sense of the word 'organ', from 'οργανον', meaning 'instrument of action' – as Bergson noted as well (CE 161). Instinct and intelligence are two modes of technical activity. They resolve the problem of how to act on matter by making and employing tools in different ways. Tools are the means by which locomotive animals modify the interface between their bodies and environments in order to navigate the world in adaptively suc-cessful ways. The locomotive regime evolves through technology.

Bergson defines arthropod instinct as the 'faculty of using and even of constructing organized instruments', and vertebrate intelligence as 'the faculty of making and using unorganized instruments' (CE 140). The distinction between instinct and intelligence as two modes of tool-use recapitulates and inverts the distinction between torpor and motility as two strategies of alimentation. Whereas plants are able to convert inor-ganic matter into nutrition, animals rely on already converted organic compounds for energy. Instinctive tool-use is like animal heterotrophy, bound to pre-specified organic matter, while intelligence is more like plant autotrophy, constructing its tools out of non-specific inorganic matter. The difference between instinct and intelligence corresponds to the difference between the types of tools that each is able to construct and employ.

The idea that organs are instruments formed by the animal in development and deployed in behaviour is independently elaborated through Raymond Ruyer's work. Bergson's idea that the human being is better considered *Homo faber* than *sapiens* finds a contemporary complement in Timothy Taylor's (2012) account of anthropogenesis via technological prostheses. The idea that instinct deploys a kind of technology of its own has garnered almost no philosophical interest. Despite the widespread appreciation for *Creative Evolution*, the theory of technology that runs through its accounts of instinct, intelligence, organ, function, knowledge, and even spatialisation has received only scant scholarly attention.[1]

4. Finally, it is now generally well-accepted that Deleuze and Guattari's philosophy of nature relies on an appreciation for the prevalence of horizontal, symbiotic co-evolutionary alliances of the sort that obtain – in Deleuze and Guattari's prime example – between the wasp and the orchid.[2] Yet there is still work to be done on the sources in philosophy and biology for that case. One of them may be Bergson's discussion of the sympathetic rapport between the wasp and the caterpillar. For all its anti-evolutionary polemics, Deleuze and Guattari's *A Thousand Plateaus* does not mark a rigid break with the interest in evolutionary theory characteristic of Bergson's work, but rather represents a path-breaking anticipation of the new directions that evolutionary theory was still then on the verge of taking. Deleuze and Guattari's discussions of the rhizomatic structure of nature and the significance of association and cross-lineage alliance over genealogical descent with modification should be thought in connection with Bergson's nascent theory of symbiosis on the one hand, and the contemporary science of endosymbiosis in phylogenesis (and as a source for new biological information) on the other. This attention to innovative work in evolutionary theory may have informed Deleuze and Guattari's other, ostensibly non-biological philosophical postulates as well.

There are a number of further avenues open for the exploration of Deleuze and Deleuze and Guattari's Bergsonian heritage. If it is granted that Bergson's metaphysics of virtuality is largely underwritten by and intended to underwrite a philosophy of biology, then Deleuze's extension of that metaphysics should be similarly conceived as well. Other philosophers of the virtual, such as Raymond Ruyer, could be incorporated in this biology-based metaphysical lineage. The result would be a rereading of a line of French metaphysicians on the basis of a primacy accorded to the life sciences in our understanding of the natural world.

★   ★   ★

If there is one hope I have for this book, it is for mine to count not only as another voice in the chorus of contemporary Bergsonism, but to offer to scholars today a powerful reason for why Bergson should be read on the basis of his engagement with Darwinism, his role as a pre-eminently evolutionary philosopher, and the resources he has to offer the philosophy of biology. In a line whose significance for the study of Bergson cannot be overstated, Bergson claimed that 'in the labyrinth of acts, states, and faculties of the mind, the thread which one must never lose is the one furnished by biology' (CM 38). I have tried to follow that thread in this book, and I hope that at least part of the contribution it makes to contemporary Bergsonism is to remind others that one of the main threads through Bergson's philosophy is the one that is furnished by biology as well.

Notes

Introduction

1. See Dupré and Nicholson 2018 (eds.) for a recent collection on these topics in the philosophy of biology.
2. There is some precedent for the focus on Bergson's philosophy of evolution. Paul-Antoine Miquel's *Le problème de la nouveauté dans l'évolution du vivant*, from 1996, is probably the best example in French, Keith Ansell-Pearson's 2002 *Philosophy and the Adventure of the Virtual* in English. There have been two recent collections of essays on *Creative Evolution* in French, *Annales bergsoniennes IV* from 2008 and L'Évolution créatrice *de Bergson* from 2010. In later work (2007b and 2014), Miquel elaborates some elements of his interpretation of Bergson's philosophy of evolution, incorporating them into a broader picture of the philosophy of science. In *Sur le concept de nature* (2015), he leverages this work into his own philosophy of nature. My approach accords in many places, though we diverge regarding the privilege afforded to intuition and imagination. Whereas Miquel conceives Bergson's thought according to the metaphysics of immanence, I prefer to centre the modality and mereology of tendency. This interpretive decision distinguishes this book from the work done in Ansell-Pearson's monograph and the edited collections as well.

Chapter 1

1. For Bergson's place in the history of theories of organic memory, see Otis 1994: 34–6
2. See Coleman 1977: 16–33 for the state of cell theory in Bergson's time.

Chapter 2

1. The reductionism of Bergson's time has since fallen out of vogue, as researchers no longer assume that memory is localised to any particular cerebral structure. Current consensus suggests that memories are encoded in the hippocampus and progressively transferred to the frontal, parietal,

and temporal lobes for longer term storage (Smith and Squire 2009). More recent accounts depart further from the classical model, as researchers now posit (somewhat confusedly designated) 'anti-memories' that work to balance, stabilise, and put offline sets of memories in order that they can be reactivated later, which allows otherwise incompatible, or impossibly large, memory associations to coexist in the same cortical structures at the same time (Barron et al. 2016).

2. Where Paul and Palmer translate '*sauter*' as 'jump' (MM 188), Lawlor substitutes the crucial 'leap' (quoted in his 2003: 41, 50). Bergson seems to prefer to speak more often of 'transporting' or 'installing' ourselves in duration, of 'adopting' and 'following' its movements, of 'penetrating' and 'seizing' the mobile objects of intuition (CM 104, 138, 156). The idea of the 'leap' – and especially the idea that it is central – was introduced into the Bergsonian lexicon by Deleuze, first in *Bergsonism* and then again in *Cinema 2*, to designate a move from out of 'psychology' and into 'ontology', or from out of the sensorimotor present and its circuit of useful memory-images and into the pure past, onto a plane of memory proper (2006: 56–7; 1989: 80). It has since been widely adopted. It is now common to read of the 'leap' in recent scholarship not only as a moment in the operation of memory, justifiable in part through translation choice, but as required by the method of intuition and so distinct from the select occasions on which Bergson may have used the word himself (see, for instance, Lefebvre 2008: 153, Grosz 2004: 179, Lawlor and Moulard-Leonard 2016: §3, Al-Saji 2004: 227).

3. See Lawlor 2020 for an analysis of the terminology of 'image', 'scheme', and 'representation' in 'Intellectual Effort'. Cf. Maras 1998 for the way this essay provides a model for actualisation beyond the domain of psychology. For Bergson's critique of the science of psychology, see Hausheer 1927.

4. At the time of 'Intellectual Effort', the phenomenon of blindfolded chess was the subject of considerable psychological attention, the definitive work of which was Binet's 1894 *Psychologie des grands calculateurs et joueurs d'écheces*. The record then for how many blind games could be played simultaneously was probably still held by Zukertort, who played sixteen games in 1876. Miguel Najdorf played forty-five games in 1947. Marc Lang set a record of forty-six games in 2011. And, most recently, in 2017, Timur Gareyev reset the record at forty-eight games.

Chapter 3

1. See Feldman 2016: 168 for more on the indexical nature of images. I am using Feldman's translation of the quoted excerpt from Bergson's letter to Delattre.

2. For the source of the term '*élan vital*' in Lalande, see Scharfstein 1942: 81; cf. Sinclair 2020: 225, n. 22.

Chapter 4

1. See Coleman 1977: 35–55 for the state of nineteeth-century biology on form and development. Cf. Wolfe 2014 for more on the history of form as an irreducible biological reality.
2. For the idea that metabolism underwrites time perception, see Posteraro 2015. There is a good deal of research on the relationship between metabolism and circadian time-keeping rhythms, especially in biochemical clocks that regulate gene expression. See Ray and Reddy 2016; cf. Hurley et al. 2014.
3. Neurophysiological research dating back to the early twentieth century suggests that the human experience of the present spans a duration of approximately 2–3 seconds. Ernst Pöppel's (1978) studies propose that humans experience these temporal intervals as units, and that this unitary perception of a present consisting of several seconds is expressed in the rhythmic iteration of everyday motor abilities. There are a number of studies that locate beneath both repetitive and non-repetitive movements in humans, as well as in other mammals, the same unified 2–3 second interval (Gerstner and Golberg 1994: 182). In Chapter 2 I suggested that the contractile limit of the perceptual second is 60 Hz. By contracting 60 elements in a second, and unifying 3 seconds in a present, our frequency profile is 180 Hz over 3 seconds. We perceive at a rate of 60 Hz in a unified present that spans 3 seconds. Research suggests that the ocelli of dragonflies have a flicker fusion threshold of 220 Hz. Their ocellar nerves are capable of perceiving just under 220 changes in a second. Beyond that threshold, changes blur and are contracted continuously. The dragonfly's organic systems beat at rates far higher than human ones. It is unsurprising that they perceive at a higher contractile rate. The dragonfly's heart rate is a typical 180 beats per minute, while that of a resting human is typically around 70. A study linking the size of an animal's body, its metabolism, and the rate at which it processes temporal information has provided some evidence for these claims (Kevin Healy et al. 2013).

Chapter 5

1. See Coleman 1977: 57–90 for the state of nineteenth-century biology on evolutionary transformation or species change. For Bergson's discussion of transformism, see François 2010: 33–6.
2. See Mullarkey 2007: 53–4 and 2008 on the *élan vital* as an explanatory injunction and not a substance of its own, part of a critical vitalism as opposed to the more traditional forms of the theory. For more on Bergson's vitalism as a heuristic, see Han 2008. See Fujita 2007 for the organ as a theme in Bergson's purportedly '(non-)organic' vitalism. For another look at Bergson's biophilosophy in the context of molecular biology, see Spassov 1998. For the history behind some of the themes of the Bergsonian vitalism, see Scharfstein 1942:

72–98. For an updated interpretation of the *élan vital*, see DiFrisco 2015. For a study of the negative function of the *élan vital*, see Caeymaex 2008. For an analysis of the status of vitality in *Creative Evolution*, see Miquel 2010: 233–8, Worms 2000: 22–3, 2004: 207–11, and 2008. Cf. Worms 2010 for the idea that Bergson's vitalism should be understood over and against nihilism. On Bergson's 'vitalism controversy', see Burwick and Douglass (eds.) 1992, and Wolsky and Wolsky 1992 in particular. For a classic overview of the history of vitalism, see Myers 1900. For more on ancient vitalism, see Federspil and Sicolo 1994. For more on the controversy between mechanism and vitalism, see Klerk 1979. For more on mechanism and vitalism in Bergson's scientific context, see Allen 2005. For more on vitalism in the life sciences, see Normandin and Wolfe (eds.) 2013. On the history of vitalism's refutation, as well as its 'return' in the context of French biophilosophy, see Wolfe and Wong 2014. For more on the controversy, see the encyclopaedia entry on vitalism in Bechtel and Richardson 1998. For a reassessment of the history of materialism, see Boyle 2018, Kaitaro 2008, Roe 1981, Reill 2005, and Wolfe 2017. For a reassessment of the history of vitalism, especially its experimental credibility, see Benton 1974, Cimino and Duchesneau (eds.) 1997, Gambarotto 2017, Smith 2011, Williams 2003, Wolfe 2011 and Wolfe and Terada 2008.

3. On the Montpellier school, see Amaral and Alfonso-Goldfarb 2011. For a critical reappraisal of their vitalism, see Wolfe and Terada 2008.

4. On this interpretation of the *élan vital*, see Deleuze 2006: 94. For a reconstruction of Deleuze's interpretation, see Lundy 2018a: 119–54. See Lapoujade 2012 for Deleuze's uptake of Bergson's concept of differentiation. See Montebello 2012b and Hwang 2008 for Deleuze's derivation of a univocal monism from Bergson's philosophy of difference.

5. See Balz 1921: 637, Barr 1913: 646, and Gunter 1999: 172. See Kreps 2015: 171–2 for a less metaphysical approach. See Ruyer 2019: 144 for a critical perspective.

6. Bergson makes similar claims, with similar qualifications, about matter as a whole and consciousness as well (MM 292–3, 313, and 331). See Dolbeault 2018 for a study of Bergson's proximity to panpsychism in this connection.

Chapter 6

1. The idea that science proceeds through the formation and modification of primarily visible images is the subject of a long and vexed debate. See Fox-Keller 1995a and 1995b, Crary 1992, Levin 1993, Jay 1993.

2. I draw from Gilbert 1991, Fox-Keller 2000: 117–19, Alseekh et al. 2017, Waddington 1956: 412. For Waddington's comments on Whitehead and philosophical biology, see his 1929 paper, 'Philosophy and Biology', qtd. in Peterson 2011: 306–7.

3. Jankélévitch has argued for a conception of 'organic possibility' in Bergson (2015: 180–1, 191). It is not clear to me what this term is intended to designate, obscured as it is by the style of Jankélévitch's writing. 'This nothingness

[of organic possibility] is now an élan toward the concrete and a *nisus forma-tivus*, now a mystical indetermination, rich and profound and sonorous like the silence of the night' (2015: 180). Gunter mentions 'organic possibility' in his article on the 'dynamic factors' that condition processes of actualisa-tion (2007: 38). It is possible that what I am calling a 'shining point' might be intended by Jankélévitch's 'organic possibility' as well. See Sinclair 2020: 190–1 for a discussion of the same idea.

4. For the research on conservation, I draw from Abouheif 2017: xvi, Bürglin and Affolter 2015, Carroll et al. 2001, Corsetti et al. 1992, Derelle et al. 2007, Hall 2007: 472–3, Hall 2012: 30, Isenbarger et al. 2008, Lowe and Wray 1997, Morris 2003: 283, Morris 2015: 136, Neumann and Nüsslein-Volhard 2000: 2139, Reneker et al. 2012, and Willmer 2003: 45.

5. For the research on developmental constraints, see Alberch 1982: 327, Amundson 1994: 570, Brigandt 2015: 306, 310, 312, Gould 2002: 1128–9, Maynard Smith et al. 1985, Parker and Maynard Smith 1990, Stephens and Krebs 1986: 180, Wagner and Larsson 2007: 50, Wesson 1991: 88–96, and Zalts and Yanai 2017.

Concluding Remarks

1. Marrati 2010 is a notable exception; Mitcham 1994 mentions Bergson's reconception of *Homo sapiens* as *Homo faber*; Ruse 2005 invokes Bergson as motivation for the philosophy of technology in general; and Steinert 2017 attempts to integrate Bergson's view on technology with his the-ory of laughter and comedy independently from the theory of technology espoused in *Creative Evolution*.

2. On Deleuze's relationship to evolutionary theory see Bennett and Posteraro (eds.) 2019.

Bibliography

Abouheif, Ehab. 2017. 'Foreword' to Lewis Held, *Deep Homology? Uncanny Similarities of Humans and Flies Uncovered by Evo-Devo*. New York: Cambridge University Press.

Alberch, Pere. 1982. 'Developmental Constraints in Evolutionary Processes'. *Evolution and Development*. Ed. J. T. Bonner. Berlin: Springer. 313–32.

Allen, Garland. 2005. 'Mechanism, Vitalism and Organicism in Late Nineteenth and Twentieth-Century Biology: The Importance of Historical Context'. *Studies in History and Philosophy of Biological and Biomedical Sciences*. Vol. 36, Issue 2: 261–83.

Allen, Frank and A. Hollenberg. 1924. 'On the Tactile Sensory Reflex'. *Quarterly Journal of Experimental Physiology*. Vol. 14, Issue 4: 351–78.

Allen, Frank and Mollie Weinberg. 1925. 'Gustatory Sensory Reflex.' *Quarterly Journal of Experimental Physiology*. Vol. 15, Issue 3: 385–420.

Al-Saji, Alia. 2004. 'The Memory of Another Past: Bergson, Deleuze, and a New Theory of Time'. *Continental Philosophy Review*. Vol. 37: 203–39.

Al-Saji, Alia. 2010. 'Life as Vision: Bergson and the Future of Seeing Differently'. *Bergson and Phenomenology*. Ed. Michael Kelly. New York: Palgrave Macmillan. 148–73.

Alseekh, Saleh, Tong, H., Scossa, F., Brotman, Y., Vigroux, F., Toge, T., Ofner, I., Zamir, D., Nikoloski, Z., and A. Fernie. 2017. 'Canalization of Fruit Metabolism'. *Plant Cell*. Vol. 29, Issue 11: 2753–65.

Amaral, Maria Thereza Cera Galvão do and Ana Alfonso-Goldfarb. 2011. 'Roots of French Vitalism: Bordeau and Barthez, between Paris and Montpellier'. Trans. Silvia Waisse. *História, Ciências, Saúde – Manguinhos*. Vol. 18, Issue 3. <https://doi.org/10.1590/S0104-59702011000300002>

Amundson, Ron. 1994. 'Two Concepts of Constraint: Adaptationism and the Challenge from Developmental Biology'. *Philosophy of Science*. Vol. 61, Issue 4: 556–78.

Anjum, Rani Lill and Stephen Mumford. 2018. *What Tends to Be: The Philosophy of Dispositional Modality*. New York: Routledge.

Ansell-Pearson, Keith. 1999. 'Bergson and Creative Evolution/Involution: Exposing the Transcendental Illusion of Organismic Life'. *The New Bergson*. Ed. John Mullarkey. New York: Manchester University Press. 146–67.

Ansell-Pearson, Keith. 2002. *Philosophy and the Adventure of the Virtual: Bergson and the Time of Life*. London: Routledge.

Ansell-Pearson, Keith. 2005a. 'The Reality of the Virtual: Bergson and Deleuze'. *Modern Language Notes*. Vol. 120, Issue 5: 1112–27.

Ansell-Pearson, Keith. 2005b. 'Bergson's Encounter with Biology'. *Angelaki*. Vol. 10, Issue 2: 59–72.

Ansell-Pearson, Keith. 2018. *Bergson: Thinking Beyond the Human Condition*. New York: Bloomsbury.

Ansell-Pearson, Keith and John Mullarkey. Eds. 2002. *Henri Bergson: Key Writings*. New York: Continuum.

Ansell-Pearson, Keith, Guerlac, Suzanne, Al-Saji, Alia, and Leonard Lawlor. 2017. 'Bergson Circle'. *56th Annual Meeting of the Society for Phenomenology and Existential Philosophy*. University of Memphis, Tennessee.

Ansell-Pearson, Keith, Miquel, Paul-Antoine, and Michael Vaughan. 2010. 'Responses to Evolution: Spencer's Evolutionism, Bergsonism, and Contemporary Biology'. *The New Century: Bergsonism, Phenomenology, and Responses to Modern Science*. Ed. Keith Ansell-Pearson and Alan D. Schrift. Durham: Acumen. 347–79.

Antliff, Mark. 1991. *Inventing Bergson: Cultural Politics and the Parisian Avant-garde*. New Jersey: Princeton University Press.

Arendt, Jeffrey and David Reznick. 2008. 'Convergence and Parallelism Reconsidered: What Have We Learned about the Genetics of Adaptation?' *Trends in Ecology and Evolution*. Vol. 23, Issue 1: 26–32.

Arlig, Andrew. 2011. 'Medieval Mereology'. *The Stanford Encyclopedia of Philosophy*. <http://plato.stanford.edu/archives/fall2011/entries/mereology-medieval/>

Ariew, Andre, Cummins, R. and M. Perlman. Eds. 2002. *Functions: New Essays in the Philosophy of Psychology and Biology*. Oxford: Oxford University Press.

Aristotle. 1984. 'Physics'. *The Complete Works of Aristotle: The Revised Oxford Translation*. Ed. Jonathan Barnes. Princeton: Princeton University Press.

Aristotle. 1995. 'De Anima'. *Selections*. Trans. Terence Irwin and Gail Fine. Indianapolis: Hackett. 169–205.

Azouvi, François. 2008. 'Le magistère bergsonien et le succès de l'élan vital'. *Annales bergsoniennes IV: L'Évolution Créatrice 1907–2007: épistémologie et métaphysique*. Ed. Frédéric Worms, Anne Fagot-Largeault, and Jean-Luc Marion. Paris: PUF. 85–93.

Bakhtin, Mikhail. 1992. 'Contemporary Vitalism'. Trans. Charles Byrd. *The Crisis of Modernism: Bergson and the Vitalist Controversy*. Ed. Frederick Burwick and Paul Douglass. New York: Cambridge University Press. 76–97.

Balz, Albert G. A. 1921. 'Reviewed Work: *Mind-Energy* by Henri Bergson, Wildon Carr'. *The Journal of Philosophy*. Vol. 18, Issue 23: 634–43.

Barr, Nann Clark. 1913. 'The Dualism of Bergson'. *The Philosophical Review*. Vol. 22, Issue 6: 639–52.

Barron, H. C., Vogels, T. P., Emir, U. E., Makin, T. R., O'Shea, J., Clare, S., Jbabdi, S., Dolan, R. J., and T. E. J. Behrens. 2016. 'Unmasking Latent Inhibitory Connections in Human Cortex to Reveal Dormant Cortical Memories'. *Neuron*. Vol. 90, Issue 1: 191–203.

Barros, Benjamin. 2008. 'Natural Selection as a Mechanism'. *Philosophy of Science*. Vol. 75: 306–22.

Beatty, John. 1995. 'The Evolutionary Contingency Thesis'. *Concepts, Theories, and Rationality in the Biological Sciences*. Ed. G. Wolters and J. G. Lennox. Pittsburgh: University of Pittsburgh Press. 45–82.

Beauregard, Oliver de. 2016. 'Bergson's Duration and Quantal Space-Time Non-Separability'. *Bergson and Modern Thought: Towards a Unified Science*. Ed. Andrew Papanicolaou and Pete Gunter. New York: Routledge. 318–42.

Bechtel, William and Robert C. Richardson. 1998. 'Vitalism'. *Routledge Encyclopedia of Philosophy*. Ed. E. Craig. London: Routledge.

Békésy, Georg. 1960. *Experiments in Hearing*. Trans. E. G. Wever. New York: McGraw-Hill.

Bennett, Jonathan. 1984. *A Study of Spinoza's* Ethics. Cambridge: Cambridge University Press.

Bennett, Michael and Tano Posteraro. 2019. 'Historical Formations and Organic Forms'. *Deleuze and Evolutionary Theory*. Ed. Michael Bennett and Tano Posteraro. Edinburgh: Edinburgh University Press. 1–22.

Benton, E. 1974. 'Vitalism in Nineteenth-Century Scientific Thought: A Typology and Reassessment'. *Studies in History and Philosophy of Science Part A*. Vol. 5, Issue 1: 17–48.

Berg, Howard. 2004. *E. Coli in Motion*. New York: Springer.

Bergman, Jerry. 2003. 'The Century-and-a-half Failure in the Quest for the Source of New Genetic Information'. *T. J. Technical Journal*. Vol. 17, Issue 2: 19–25.

Bergson, Henri. 1920. *Mind-Energy*. Trans. H. Wildon Carr. New York: Henry Holt and Co.

Bergson, Henri. 1972. *Mélanges*. Ed. André Robinet. Paris: PUF.

Bergson, Henri. 1977. *The Two Sources of Morality and Religion*. Trans. R. Ashley Audra and Cloudesley Brereton. Indiana: University of Notre Dame Press.

Bergson, Henri. 1990. *Cours I: Leçons de psychologie et de métaphysique*. Ed. Henri Hude. Paris: PUF.

Bergson, Henri. 1998. *Creative Evolution*. Trans. Arthur Mitchell. Mineola: Dover.

Bergson, Henri. 2001. *Time and Free Will: An Essay on the Immediate Data of Consciousness*. Trans. F. L. Pogson. Mineola: Dover.

Bergson, Henri. 2002. *Correspondences*. Paris: PUF.

Bergson, Henri. 2004. *Matter and Memory*. Trans. Nancy M. Paul and W. Scott Palmer. Mineola: Dover.

Bergson, Henri. 2007. *The Creative Mind: An Introduction to Metaphysics*. Trans. Mabelle L. Andison. Mineola: Dover.

Bergson, Henri. 2007. 'Cours sur le *De rerum originatione radicali* de Leibniz'. Ed. Matthias Vollet. *Annales bergsoniennes III: Bergson et la Science*. Ed Frédéric Worms. Paris: PUF. 35–52.

Bergson, Henri. 2017. *L'évolution du problème de la liberté. Cours au Collège de France 1904–1905*. Ed. Arnaud François. Paris: PUF.

Bergson, Henri. 2018. *Histoire des théories de la mémoire. Cours au Collège de France 1903–1904*. Ed. Arnaud François. Paris: PUF.

Bergson, Henri and Floris Delattre. 1936. 'Samuel Butler et le bergsonisme. Avec deux lettres inédites d'Henri Bergson'. *Revue anglo-américaine*. Vol. 8, Issue 5: 385–405.

Bernet, Rudolf. 2010. 'Bergson on the Driven Force of Consciousness and Life'. *Bergson and Phenomenology*. Ed. Michael Kelly. New York: Palgrave Macmillan. 42–62.

Bernet, Rudolf. 2012. 'La conscience et la vie comme force et pulsion'. *Bergson*. Ed. Camille Riquier. Paris: Cerf. 25–54.

Bertalanffy, Karl Ludwig. 1932. *Theoretische Biologie, Band 1: Allgemeine Theorie, Physikochemie, Aufbau und Entwicklung des Organismus*. Berlin: Gebrüder Bortraeger.

Bertalanffy, Karl Ludwig. 1950. 'The Theory of Open Systems in Physics and Biology'. *Science*. Vol. 111, Issue 2872: 23–9.

Bertalanffy, Karl Ludwig. 1968. *General System Theory*. New York: George Braziller.

Berthoz, Alain. 2008. 'Les théories de Bergson sur la perception, la mémoire et le rire, au regard des données des neurosciences'. *Annales bergsoniennes IV: L'Évolution créatrice 1907–2007: épistémologie et métaphysique*. Ed. Frédéric Worms, Anne Fagot-Largeault, and Jean-Luc Marion. Paris: PUF. 163–78.

Bianco, Giuseppe. 2011. 'Experience vs. Concept? The Role of Bergson in Twentieth-Century French Philosophy'. *The European Legacy: Toward New Paradigms.* Vol. 16, Issue 7: 855–72.

Bianco, Giuseppe. 2019. 'Philosophies of Life'. *The Cambridge History of Modern European Thought.* Ed. P. Gordon and W. Breckman. Cambridge: Cambridge University Press. 153–75.

Bing, Franklin. 1971. 'The History of the Word "Metabolism"'. *Journal of the History of Medicine and Allied Sciences.* Vol. 26, Issue 2: 158–80.

Blanchard, Pascal. 2008. 'La métaphysique de la matière'. *Annales bergsoniennes IV: L'Évolution créatrice 1907–2007: épistémologie et métaphysique.* Ed. Frédéric Worms, Anne Fagot-Largeault, and Jean-Luc Marion. Paris: PUF. 499–512.

Bognon, Cécilia, Chen, Bohang, and Charles Wolfe. 2018. 'Metaphysics, Function and the Engineering of Life: The Problem of Vitalism'. *Kairos.* Vol. 20, Issue 1: 113–40.

Bohm, David. 2002. *Wholeness and the Implicate Order.* New York: Routledge.

Bowler, Peter J. 1979. 'Theodor Eimer and Orthogenesis: Evolution by "Definitely Directed Variation"'. *Journal of the History of Medicine and Allied Sciences.* Vol. 36, Issue 1: 40–73.

Bowler, Peter J. 1992. *The Eclipse of Darwin: Anti-Darwinian Evolution Theories in the Decades around 1900.* Baltimore: Johns Hopkins University Press.

Boyle, Deborah. 2018. *The Well-Ordered Universe: The Philosophy of Margaret Cavendish.* Oxford University Press.

Bradley, F. H. 1893. *Appearance and Reality.* London: George Allen & Unwin.

Bray, Dennis. 2009. *Wetware: A Computer in Every Living Cell.* New Haven: Yale University Press.

Brigandt, Ingo. 2015. 'From Developmental Constraint to Evolvability: How Concepts Figure in Explanation and Disciplinary Identity'. *Conceptual Change in Biology: Scientific and Philosophical Perspectives on Evolution and Development.* Ed. Alan C. Love. New York: Springer. 305–26.

Bürglin, T. R. and M. Affolter. 2015. 'Homeodomain Proteins: An Update'. *Chromosoma.* Vol. 125, Issue 3: 1–25.

Burton, James. 2008. 'Bergson's Non-Archival Theory of Memory'. *Memory Studies.* Vol. 1, Issue 3: 321–39.

Burwick, Frederick and Paul Douglass. Eds. 1992. *The Crisis in Modernism: Bergson and the Vitalist Controversy.* Cambridge: Cambridge University Press.

Butterfield, Jeremy. 2011. 'Less is Different: Emergence and Reduction Reconciled'. *Foundations of Physics.* Vol. 41: 1065–135.

Caeymaex, Florence. 2008. 'Négativité et finitude de l'élan vital. La lecture de Bergson par Jankélévitch'. *Annales bergsoniennes IV: L'Évolution créatrice 1907–2007: épistémologie et métaphysique.* Ed. Frédéric Worms, Anne Fagot-Largeault, and Jean-Luc Marion. Paris: PUF. 629–40.

Caeymaex, Florence. 2013. 'The Comprehensive Meaning of Life in Bergson'. Trans. Edward F. McGushin. *The Science, Politics, and Ontology of Life-Philosophy.* Ed. Scott M. Campbell and Paul W. Bruno. New York: Bloomsbury. 47–64.

Caillois, Roger. 1976. *Coherence aventureuses.* Paris: Gallimard.

Campbell, Donald. 1974. 'Downward Causation in Hierarchically Organised Biological Systems'. *Studies in the Philosophy of Biology.* Ed. F. J. Ayala and T. Dobzhansky. London: Palgrave Macmillan. 179–86.

Canales, Jimena. 2015. *The Physicist and the Philosopher: Einstein, Bergson, and the Debate that Changed our Understanding of Time.* Princeton: Princeton University Press.

Canguilhem, Georges. 2002. 'Le concept et la vie'. *Études d'histoire et de philosophie des sciences*, 7th ed. Paris: Vrin.

Canguilhem, Georges. 2008. *Knowledge of Life.* Trans. Stefanos Geroulanos and Daniela

Ginsburg. Ed. Paolo Marrati and Todd Meyers. New York: Fordham University Press.

Čapek, Milič. 1971. *Bergson and Modern Physics: A Reinterpretation and Re-evaluation.* Dordrecht: D. Reidel.

Čapek, Milič. 2016. 'Bergson's Theory of the Mind-Brain Relation'. *Bergson and Modern Thought: Towards a Unified Science.* Ed. Andrew Papanicolaou and Pete Gunter. New York: Routledge. 129–48.

Carroll, S. B., Grenier, J. K., and S. D. Weatherbee. 2001. *From DNA to Diversity: Molecular Genetics and the Evolution of Animal Design.* Malden: Blackwell.

Cartwright, Nancy. 1999. *The Dappled World.* Cambridge: Cambridge University Press.

Cartwright, Nancy. 2007. 'What Makes a Capacity a Disposition?' *Dispositions and Causal Powers.* Ed. Max Kistler and Bruno Gnassounou. Aldershot: Ashgate.

Carvalho, Magda Costa. 2008. 'La biologie et la psychologie: les "clefs de contact" du dynamism vitaliste bergsonien'. *Annales bergsoniennes IV: L'Évolution créatrice 1907–2007: épistémologie et métaphysique.* Ed. Frédéric Worms, Anne Fagot-Largeault, and Jean-Luc Marion. Paris: PUF. 539–48.

Carvalho, Magda Costa and M. Patrão Neves. 2010. 'Building the "True Evolutionism": Darwin's Impact on Henri Bergson's Thought'. *Revista Portuguesa de Filosofia, Evolução, Ética e Cultura.* Vol. 66, Issue 3: 635–42.

Cesari, P., Formenti, F. and P. Olivato. 2003. 'A Common Perceptual Parameter for Stair Climbing for Children, Young and Old Adults'. *Human Movement Science* 1: 111–24.

Charlesworth, Brian and Deborah Charlesworth. 2009. 'Darwin and Genetics'. *Genetics*. Vol. 183, Issue 3: 757–66.

Chemero, Anthony. 2009. *Radical Embodied Cognitive Science*. Cambridge: MIT.

Chen, Bohang. 2018. 'A Non-Metaphysical Evaluation of Vitalism in the Early Twentieth Century'. *History and Philosophy of the Life Sciences*. Vol. 40, Issue 3. <https://doi.org/10.1007/s40656–018–0221–2>

Chimisso, Cristina. 2008. *Writing the History of the Mind: Philosophy and Science in France, 1900 to 1960s*. Burlington: Ashgate.

Cimino, Guido and François Duchesneau. Eds. 1997. *Vitalisms: From Haller to the Cell Theory*. Florence: Leo Olschki.

Cohen, Richard. 1999. 'Philo, Spinoza, Bergson: The Rise of an Ecological Age'. *The New Bergson*. Ed. John Mullarkey. New York: Manchester University Press. 18–31.

Coleman, William. 1977. *Biology in the Nineteenth Century: Problems of Form, Function, and Transformation*. New York: Cambridge University Press.

Conger, George P. 1927. 'Whitehead Lecture Notes: Logical and Metaphysical Problems'. *Manuscripts and Archives*. Yale University Library. New Haven, Connecticut.

Cook, Nicholas. 2016. *Music, Performance, Meaning*. New York: Routledge.

Cope, Edward. 1896. *The Primary Factors of Organic Evolution*. Chicago: Open Court.

Corsetti, M. T., Briata, P., Sanseverino, L., Daga, A., Airoldi, I., Simeone, A., Palmisano, G., Angelini, C., Boncinelli, aE., and G. Corte. 1992. 'Differential DNA Binding Properties of Three Human Homeodomain Proteins'. *Nucleic Acids Research*. Vol. 20, Issue 17: 4465–72.

Costelloe, Karin. 1913. 'What Bergson Means by "Interpenetration"'. *Proceedings of the Aristotelian Society*. Vol. 13: 131–55.

Cowling, S. 2014. 'Instantiation as Location'. *Philosophical Studies*. Vol. 167: 667–82.

Crary, Jonathan. 1992. *Techniques of the Observer: On Vision and Modernity in the Nineteenth Century*. Cambridge: MIT.

Crick, Francis H. 1958. 'On Protein Synthesis'. *Symposia of the Society for Experimental Biology*. Vol. 12: 138–63.

Cunningham, G. Watts. 1914. 'Bergson's Conception of Finality'. *The Philosophical Review*. Vol. 23, Issue 6: 648–63.

Darwin, Charles. 1868. *The Variation of Animals and Plants under Domestication*. London: John Murray.

Darwin, Charles. 1960. 'Darwin's Notebooks on Transmutation of Species: Part IV, Fourth Notebook'. Ed. Gavin de Beer. *Bulletin of the British Museum (Natural History)*. Vol. 2, Issue 5: 151–83.

Darwin, Charles. 1984. *The Various Contrivances by which Orchids are Fertilised by Insects*. 2nd. ed. Chicago: University of Chicago Press.

Darwin, Charles. 2003. *The Origin of Species*. New York: Penguin.

Dawkins, Richard. 1989. *The Selfish Gene*. Oxford: Oxford University Press.

Debaise, Didier. 2009. 'The Emergence of a Speculative Empiricism: Whitehead Reading Bergson'. *Deleuze, Whitehead, Bergson: Rhizomatic Connections*. Ed. Keith Robinson. New York: Palgrave Macmillan. 77–88.

Debaise, Didier. 2017. *Nature as Event: The Lure of the Possible*. Trans. Michael Halewood. Durham: Duke University Press.

DeLanda, Manuel. 2002. *Intensive Science and Virtual Philosophy*. London: Continuum.

Deleuze, Gilles. 1989. *Cinema 2: The Time-Image*. Trans. Hugh Tomlinson and Robert Galeta. London: The Athlone Press.

Deleuze, Gilles. 1991. *Empiricism and Subjectivity: An Essay on Hume's Theory of Human Nature*. Trans. Constantin V. Boundas. New York: Columbia University Press.

Deleuze, Gilles. 1994. *Difference and Repetition*. Trans. Paul Patton. New York: Columbia University Press.

Deleuze, Gilles. 2004. *Desert Islands and Other Texts 1953–1974*. Ed. David Lapoujade. Trans. Michael Taormina. Los Angeles: Semiotext(e).

Deleuze, Gilles. 2006. *Bergsonism*. Trans. Hugh Tomlinson and Barbara Habberjam. New York: Zone Books.

Deleuze, Gilles. 2007. 'Theory of Multiplicities in Bergson'. *Lectures by Gilles Deleuze*. <http://deleuzelectures.blogspot.com/2007/02/theory-of-multiplicities-in-bergson.html>

Dennett, Daniel. 1995. *Darwin's Dangerous Idea: Evolution and the Meanings of Life*. London: Allen Lane.

Derelle, R., Lopez, P., Le Guyader, H., and M. Manuel. 2007. 'Homeodomain Proteins Belong to the Ancestral Molecular Toolkit of Eukaryotes'. *Evolution & Development*. Vol. 9, Issue 3: 212–19.

Devarieux, Anne. 2012. 'Maine de Biran – Henri Bergson. L'avenir de la volonté'. *Bergson*. Ed. Camille Riquier. Paris: Cerf. 163–89.

DiFrisco, James. 2015. '*Élan Vital* Revisited: Bergson and the Thermodynamic Paradigm'. *The Southern Journal of Philosophy*. Vol. 53, Issue 1: 54–73.

Dobzhansky, Theodosius. 1967. 'L'Évolution créatrice'. *Diogenes* 58: 64–80.

Dolbeault, Joël. 2008. 'L'idée que la matière a une mémoire, dans *L'Évolution créatrice*'. *Annales bergsoniennes IV: L'Évolution Créatrice 1907–2007: épistémologie et métaphysique*. Ed. Frédéric Worms, Anne Fagot-Largeault, and Jean-Luc Marion. Paris: PUF. 529–37.

Dolbeault, Joël. 2018. 'Bergson's Panpsychism'. *Continental Philosophy Review*. Vol. 51: 549–64.

Driesch, Hans. 1914. *The History and Theory of Vitalism*. Trans. C. K. Ogden. London: Macmillan.

Driesch, Hans. 1929. *The Science and Philosophy of the Organism*. 2nd ed. London: Adam and Charles Black.

Dunham, Jeremy. 2016. 'A Universal and Absolute Spiritualism: Maine de Biran's Leibniz'. *Maine de Biran: The Relationship of the Physical and the Moral in Man*. Ed. Darian Meacham and Joseph Spadola. London: Bloomsbury. 157–92.

Duns Scotus, John. 2000. *A Treatise on Potency and Act. Questions on the Metaphysics of Aristotle. Book IX*. Ed. Allan Wolter. St. Bonaventure: Franciscan Institute.

Dupré, John. 2017. 'The Metaphysics of Evolution'. *Interface Focus*. Vol. 7, Issue 5: 20160148.

Dupré, John and Dan Nicholson. Eds. 2018. *Everything Flows: Towards a Processual Philosophy of Biology*. New York: Oxford University Press.

Dupré, John and Maureen A. O'Malley. 2009. 'Varieties of Living Things: Life at the Intersection of Lineage and Metabolism'. *Philosophy & Theory in Biology*. Vol. 1: 1–25.

During, Eli. 2004. '"A History of Problems": Bergson and the French Epistemological Tradition'. *Journal of the British Society for Phenomenology*. Vol. 35, Issue 1: 4–23.

Eimer, Theodor Gustav Heinrich. 1898. *On Orthogenesis and the Impotence of Natural Selection in Species Formation*. Trans. J. M. McCormack. Chicago: Open Court.

Epperson, Michael. 2004. *Quantum Mechanics and the Philosophy of Alfred North Whitehead*. New York: Fordham University Press.

Espenson, J. H. 2002. *Chemical Kinetics and Reaction Mechanisms*. 2nd ed. New York: McGraw-Hill.

Faraday, Michael. 1844. 'A Speculation Touching Electric Conduction and the Nature of Matter'. *Philosophical Magazine*. Vol. 24: 136–44.

Federspil, Giovanni and Nicola Sicolo. 1994. 'The Nature of Life in the History of Medical and Philosophic Thinking'. *Journal of the American Society of Nephrology* 14: 337–43.

Feldman, Alex. 2016. 'The Concept in Life and the Life of the Concept: Canguilhem's Final Reckoning with Bergson'. *Journal of French and Francophone Philosophy*. Vol. 24, No. 2: 154–75.

Fox-Keller, Evelyn. 1995a. *Refiguring Life: Metaphors of Twentieth-Century Biology*. New York: Columbia University Press.

Fox-Keller, Evelyn. 1995b. 'The Biological Gaze'. *FutureNatural: Nature, Science, Culture*. Ed. G. Robertson, M. Mash, L. Tickner, J. Bird, B. Curtis, and T. Putnam. New York: Routledge.

Fox-Keller, Evelyn. 2000. *The Century of the Gene*. Cambridge: Harvard University Press.

François, Arnaud. 2008. 'Les sources biologiques de *L'Évolution créatrice*'. *Annales bergsoniennes IV: L'Évolution créatrice 1907–2007: épistémologie et métaphysique*. Ed. Frédéric Worms, Anne Fagot-Largeault, and Jean-Luc Marion. Paris: PUF. 95–109.

François, Arnaud. Ed. 2010. L'Évolution créatrice *de Bergson. Études et commentaires*. Paris: Vrin.

Frank, Philip. 1998. *The Law of Causality and its Limits*. Ed. Marie Neurath and Robert Cohen. Dordrecht: Kluwer.

Freud, Sigmund. 1961. *Beyond the Pleasure Principle*. Trans. James Strachey. New York: Liveright.

Freyhofer, Horst Heinz. 1982. *The Vitalism of Hans Driesch: The Success and Decline of a Scientific Theory*. Frankfurt: Peter Lang Gmbh.

Fujita, Hisashi. 2007. 'Bergson's Hand: Toward a History of (Non)-Organic Vitalism'. *SubStance*. Vol. 36, Issue 3: 115–30.

Gambarotto, Andrea. 2017. *Vital Forces, Teleology and Organization: Philosophy of Nature and the Rise of Biology in Germany*. Cham: Springer.

Garrett, Brian. 2013. 'Vitalism Versus Emergent Materialism'. *Vitalism and the Scientific Image in Post-Enlightenment Life Science, 1800–2010*. Ed. Sebastian Normandin and Charles Wolfe. New York: Springer. 127–54.

Gaskell, Walter. 1886. 'On the Structure, Distribution, and Functions of the Nerves which Innervate the Visceral and Vascular Systems'. *The Journal of Physiology*. Vol. 7, Issue 1: 1–80.

Gayon, Jean. 1998. 'La philosophie et la biologie'. *Encyclopédie philosophique universelle*. Vol. IV. Ed. J.-F. Mattéi. Paris: PUF. 2152–71.

Gayon, Jean. 2005. 'Bergson's Spiritualist Metaphysics and the Sciences'. *Continental Philosophy of Science*. Ed. Gary Gutting. Malden: Blackwell. 43–58.

Gayon, Jean. 2006. 'Hérédité des caractères acquis'. *Lamarck, Philosophe de la Nature*. Ed. P. Corsi, J. Gayon, G. Gohau, and S. Tirard. Paris: PUF. 105–63.

Gayon, Jean. 2008. '*L'Évolution créatrice* lue par les fondateurs de la théorie synthétique de l'évolution'. *Annales Bergsoniennes IV: L'Évolution créatrice 1907–2007: épistémologie et métaphysique*. Ed. Frédéric Worms, Anne Fagot-Largeault, and Jean-Luc Marion. Paris: PUF. 59–93.

Gehring, Walter. 1996. 'The Master Control Gene for Morphogenesis and Evolution of the Eye'. *Genes to Cells*. Vol. 1, Issue 1: 11–15.

Gehring, Walter and Makiko Seimiya. 2010. 'Eye Evolution and the Origin of Darwin's Eye Prototype'. *Italian Journal of Zoology*. Vol. 77, Issue 2: 124–36.

Gerhart, John and Marc Kirschner. 1997. *Cells, Embryos, and Evolution*. Oxford: Blackwell.

Gerstner, Geoffrey and Louis Goldberg. 1994. 'Evidence of a Time Constant Associated with Movement Patterns in Six Mammalian Species'. *Ethology and Sociobiology*. Issue 15: 181–205.

Gibbs, Sarah. 1978. 'The Chloroplasts of *Euglena* May Have Evolved from Symbiotic Green Algae'. *Canadian Journal of Botany*. Vol. 56, Issue 22: 2883–9.

Gibson, J. J. 1979. *Ecological Approach to Visual Perception*. Boston: Houghton Mifflin.

Gilbert, Scott. 1982. 'Intellectual Traditions in the Life Sciences: Molecular Biology and Biochemistry'. *Perspectives in Biology and Medicine*. Vol. 26: 151–62.

Gilbert, Scott. 1991. *A Conceptual History of Modern Embryology*. New York: Plenum.

Gitelson, I., Lisvosky, G. M., and R. D. MacElroy. 2003. *Manmade Closed Ecological Systems*. New York: Taylor & Francis.

Glynn, Luke. 2012. '*Getting Causes from Powers*, by Stephen Mumford and Rani Lill Anjum'. Review. *Mind*. Vol. 121, Issue 484: 1099–106.

Goddard, Jean-Christophe. 2010. 'La science modern et la métaphysique de la vie dans le chaptire IV de *L'Évolution créatrice*'. L'Évolution créatrice *de Bergson. Études et commentaires*. Ed. Arnaud François. Paris: Vrin. 307–20.

Godfrey-Smith, Peter. 1996. *Complexity and the Function of Mind in Nature*. New York: Cambridge University Press.

Godfrey-Smith, Peter. 1999. 'Adaptationism and the Power of Selection'. *Biology and Philosophy*. Vol. 14, Issue 2: 181–94.

Goodwin, Brian. 1994. *How the Leopard Changed its Spots: The Evolution of Complexity*. New York: Simon & Schuster.

Gould, Stephen Jay. 1977. 'Eternal Metaphors of Palaeontology'. *Patterns of Evolution as Illustrated by the Fossil Record*. Ed. A. Hallam. Amsterdam: Elsevier. 1–26.

Gould, Stephen Jay. 2001. 'Punctuated Equilibrium'. *Darwin: A Norton Critical Edition*. Ed. Philip Appleman. New York: W. W. Norton & Company. 344–9.

Gould, Stephen Jay. 2002. *The Structure of Evolutionary Theory*. Cambridge: Belknap.

Gould, Stephen Jay and Niles Eldridge. 2004. 'Punctuated Equilibria: An Alternative to Phyletic Gradualism'. *Evolution*. Ed. Mark Ridley. New York: Oxford University Press. 82–115.

Gould, Stephen Jay and Richard Lewontin. 2006. 'The Spandrels of San Marco and the Panglossian Paradigm: A Critique of the Adaptationist Programme'. *Conceptual Issues in Evolutionary Biology*. 3rd ed. Ed. Elliott Sober. Cambridge: MIT. 79–98.

Green, Sara. 2007. 'Scale Dependency and Downward Causation in Biology'. *Philosophy of Science*. Vol. 85, Issue 5: 998–1011.

Griffiths, A. J. F., Miller, J. H., Suzuki, D. T. et al. 2000. *An Introduction to Genetic Analysis*. 7th ed. New York: W. H. Freeman.

Grosso, Michael. 2015. 'The "Transmission" Model of Mind and Body: A Brief History'. *Beyond Physicalism: Toward Reconciliation of Science and Spirituality*. Ed. Edward Kelly, Adam Crabtree, and Paul Marshall. New York: Rowman & Littlefield. 79–114.

Grosz, Elizabeth. 2004. *The Nick of Time: Politics, Evolution, and the Untimely*. Durham: Duke University Press.

Grosz, Elizabeth. 2007. 'Deleuze, Bergson and the Concept of Life'. *Revue internationale de philosophie*. Vol. 3, No. 241: 287–300.

Gruner, Stefan. 2017. 'Hans Driesch Re-Visited After a Century: On "Leib und Seele – Eine Untersuchung Über das Psychophysische Grundproblem"'. *Cosmos and History*. Vol. 13, Issue 3: 401–24.

Guerlac, Suzanne. 2006. *Thinking in Time: An Introduction to Henri Bergson*. Ithaca: Cornell University Press.

Gunter, Pete. 1982. 'Bergson and Jung'. *Journal of the History of Ideas*. Vol. 43, Issue 4: 635–52.

Gunter, Pete. 1991. 'Bergson and Non-linear Non-equilibrium Thermo-dynamics: An Application of Method'. *Revue Internationale de Philosophie*. Vol. 45: 108–21.

Gunter, Pete. 1999. 'Bergson and the War against Nature'. *The New Bergson*. Ed. John Mullarkey. New York: Manchester University Press. 168–83.

Gunter, Pete. 2007. 'Bergson's Creation of the Possible'. *SubStance*. Vol. 36, Issue 3: 33–41.

Halder, Georg, Callaerts, Patrick, and Walter Gehring. 1995. 'Induction of Ectopic Eyes by Targeted Expression of the Eyeless Gene in *Drosophila*'. *Science*. Vol. 267, Issue 5205: 1788–92.

Hall, Brian K. 2007. 'Tapping Many Sources: The Adventitious Roots of Evo-Devo in the Nineteenth Century'. *From Embryology to Evo-Devo: A History of Developmental Evolution*. Ed. Manfred Laubichler and Jane Maienschein. Cambridge: MIT. 467–98.

Hall, Brian K. 2012. 'Parallelism, Deep Homology, and Evo-Devo'. *Evolution & Development*. Vol. 14, Issue 1: 29–33.

Han, Hee-Jin. 2008. 'L'heuristique du vitalisme: le principe vital de barthez et l'élan vital de Bergson'. *Annales bergsoniennes IV: L'Évolution créatrice 1907–2007: épistémologie et métaphysique*. Ed. Frédéric Worms, Anne Fagot-Largeault, and Jean-Luc Marion. Paris: PUF. 147–61.

Hartz, Glenn. 2011. 'Leibniz's Animals: Where Teleology Meets Mechanism'. *Machines of Nature and Corporeal Substances in Leibniz*. Ed. Justin Smith and Ohad Nachtomy. New York: Springer. 29–38.

Hausheer, Herman. 1927. 'Bergson's Critique of Scientific Psychology'. *The Philosophical Review*. Vol. 35, Issue 5: 450–61.

Havstad, Joyce. 2011. 'Discussion: Problems for Natural Selection as a Mechanism'. *Philosophy of Science*. Vol. 78: 512–23.

Healy, Kevin et al. 2013. 'Metabolic Rate and Body Size are Linked with Perception of Temporal Information'. *Animal Behaviour*. Vol. 86, Issue. 4: 685–96.

Hearst, Eliot and John Knott. 2009. *Blindfold Chess: History, Psychology, Techniques, Champions, World Records, and Important Games*. Jefferson: McFarland.

Hegel, G. W. F. 1969. *Hegel's Science of Logic*. Trans. A. V. Miller. London: George Allen & Unwin.

Hecht, S. and E. Smith. 1936. 'Intermittent Stimulation by Light: Area and the Relation Between Critical Frequency and Intensity'. *Journal of General Physiology*. Vol. 19, Issue 6: 979–89.

Heidel, Andrew, Barazani, Oz, and Ian Baldwin. 2010. 'Interaction Between Herbivore Defense and Microbial Signaling: Bacterial Quorum-Sensing Compounds Weaken JA-Mediated Herbivore Resistance in Nicotiana Attenuate'. *Chemoecology*. Vol. 20: 149–54.

Heil, Martin and Richard Karban. 2009. 'Explaining Evolution of Plant Communication by Airborne Signals'. *Trends in Ecology and Evolution*. Vol. 25, Issue 3: 137–44.

Heim, Noel, Knope, M., Schaal, E., Wang, S., and Jonathan Payne. 2015. 'Cope's Rule in the Evolution of Marine Animals'. *Science*. Vol. 347, Issue 6224: 867–70.

Helm, Bertrand P. 1985. *Time and Reality in American Philosophy*. Amherst: University of Massachusetts.

Helmholtz, Hermann von. 1873. *Popular Lectures on Scientific Subjects*. Trans. E. Atkinson. New York: Appleton.

Hempel, Carl. 1965. 'Studies in the Logic of Explanation'. *Aspects of Scientific Explanation, and Other Essays in the Philosophy of Science*. New York: Free Press.

Herring, Emily. 2018. '"Great is Darwin and Bergson his Poet": Julian Huxley's Other Evolutionary Synthesis'. *Annals of Science*. Vol. 75, Issue 1: 40–54.

Hirai, Yasushi. 2018. 'Event and Mind: An Expanded Bergsonian Perspective'. *Colloquium on Understanding Digital Events*. June. 1–17.

Hull, David L. 1981. 'Philosophy and Biology'. *Contemporary Philosophy: A New Survey*. Ed. G. Fløistad. The Hague: Martinus Nijhoff. 281–316.

Hurley, J., Dasgupta, A., Emerson, J., Zhou, X., Ringelberg, C., Knabe, N., Lipzen, A., Lindquist, E., Daum, C., Barry, K., Grigoriev, I., Smith, K., Galagan, J., Bell-Pedersen, D., Freitag, M., Cheng, C., Loros, J., Dunlap, J. 2014. 'Analysis of Clock-regulated Genes in *Neurospora* Reveals Widespread Posttranscriptional Control of Metabolic Potential'. *PNAS*. Vol. 111, Issue 48: 16995–17002.

Hwang, Su-Young. 2008. 'Le monism de la difference. Examen de l'interprétation deleuzienne de Bergson'. *Annales bergsoniennes IV: L'Évolution créatrice 1907–2007: épistémologie et métaphysique*. Ed. Frédéric Worms, Anne Fagot-Largeault, and Jean-Luc Marion. Paris: PUF. 655–66.

Isenbarger, T. et al. 2008. 'The Most Conserved Genome Segments for Life Detection on Earth and Other Planets'. *Origins of Life and Evolution of Biospheres*. Vol. 38, Issue 6: 517–33.

Jacob, François. 1997. 'Evolution and Tinkering'. *Science*. Vol. 196, Issue 4295: 1161–6.

Jacob, François and Jacques Monod. 1961. 'Genetic Regulatory Mechanisms in the Synthesis of Proteins'. *Journal of Molecular Biology*. Vol. 3, Issue 3: 318–56.

Jankélévitch, Vladimir. 2015. *Henri Bergson*. Ed. Alexandre Lefebvre. Trans. Nils Schott. Durham: Duke University Press.

Jay, Martin. 1993. *Downcast Eyes: The Denigration of Vision in Twentieth-Century French Thought*. Berkeley: California University Press.

Jorati, Julia. 2017. *Leibniz on Causation and Agency*. New York: Cambridge University Press.

Kaitaro, Timo. 2008. 'Can Matter Mark the Hours? Eighteenth-Century Vitalist Materialism and Functional Properties'. *Science in Context*. Vol. 21, Issue 4: 581–92.

Kanamori, Osamu. 2010. 'Chapitre Premier. L'évolution de la vie. Mécanisme et finalité'. *L'Évolution créatrice de Bergson. Études et commentaires*. Ed. Arnaud François. Paris: Vrin. 111–23.

Kant, Immanuel. 2000. *Critique of the Power of Judgment*. Trans. Paul Guyer. Ed. Allen Wood Cambridge: Cambridge University Press.

Karban, Richard. 2015. *Plant Sensing and Communication*. London: University of Chicago Press.

Karban, Richard, Baldwin, Ian, Baxter, K., Laue, G., and G. Felton. 2000. 'Communication Between Plants: Induced Resistance in Wild Tobacco Plants Following Clipping of Neighboring Sagebrush'. *Oecologia*. Vol. 125, Issue 1: 66–71.

Kauffman, Stuart. 1995. *At Home in the Universe: The Search for the Laws of Self-Organization and Complexity*. New York: Oxford University Press.

Khandker, Wahinda. 2014. 'Life as Method: The Invention of Problems in Deleuze's *Bergsonism*'. *Understanding Deleuze, Understanding Modernism*. Ed. Paul Ardoin, S. E. Gontarski, and Laci Mattison. New York: Bloomsbury. 21–32.

Kirschner, Marc. 2013. 'Beyond Darwin: Evolvability and the Generation of Novelty'. *BMC Biology*. Vol. 11, no. 110. <https://doi.org/10.1186/1741–7007–11–110>

Kistler, Max and Bruno Gnassounou. Eds. 2007. *Dispositions and Causal Powers*. Aldershot: Ashgate.

Klaus, Sander. 1997. *Entelechy and the Ontogenetic Machine: Work and Views of Hans Driesch from 1895 to 1910*. Berlin: Springer.

Klerk, Geert-Jan de. 1979. 'Mechanism and Vitalism: A History of the Controversy'. *Acta Biotheoretica*. Vol. 28, Issue 1: 1–10.

Kolkman, Michael and Michael Vaughan. Eds. 2007. 'Henri Bergson's *Creative Evolution* 100 Years Later'. *SubStance*. Vol. 36, No. 3, Issue 114.

Kreps, David. 2015. *Bergson, Complexity, and Creative Emergence*. New York: Palgrave Macmillan.

Lamarck, Jean Baptiste. 1801. *Système des animaux sans vertèbres. . .; précédé du discours d'ouverture de l'an VIII de la République*. Paris: Déterville.

Lamarck, Jean Baptiste. 1802. *Recherches sur l'organisation des corps vivans . . .; précédé du discours d'ouverture du course de zoologie, donné dans le Muséum national d'Histoire Naturelle*. Paris: Maillard.

Lamb, Marion and Eva Jablonka. 2005. *Evolution in Four Dimensions: Genetic, Epigenetic, Behavioral, and Symbolic Variation in the History of Life*. Cambridge: MIT.

Landecker, Hannah. 2013. 'Metabolism, Reproduction, and the Aftermath of Categories'. *The Scholar and Feminist Online*. 'Life (Un)ltd: Feminism, Bioscience, Race'. Ed. Rachel Lee. Vol. 11, Issue 3. <http://sfonline.barnard.edu/life-un-ltd-feminism-bioscience-race/metabolism-reproduction-and-the-aftermath-of-categories/0/>

Lapoujade, David. 2008. 'Sur un concept méconnu de Bergson: L'attachement à la vie. Pour une relecture des *Deux Sources de la Morale et de la Religion*'. *Annales bergsoniennes IV: L'Évolution Créatrice 1907–2007: épistémologie et métaphysique*. Ed. Frédéric Worms, Anne Fagot-Largeault, and Jean-Luc Marion. Paris: PUF. 673–93.

Lapoujade, David. 2012. '1956: Deleuze enveloppe Bergson'. *Bergson*. Ed. Camille Riquier. Paris: Cerf. 239–49.

Larson, Edward. 2004. *Evolution: The Remarkable History of a Scientific Theory*. New York: The Modern Library.

Latour, Bruno. 2004. 'How to Talk About the Body? The Normative Dimension of Science Studies'. *Body & Society*. Vol. 10, Issue 2: 205–29.

Lawlor, Leonard. 2003. *The Challenge of Bergsonism: Phenomenology, Ontology, Ethics*. London: Continuum.

Lawlor, Leonard. 2018. '"Machine à contingence": Bergson's Theory of Freedom in *L'évolution du problème de la liberté*'. *Lo Sguardo* 26: 47–64.

Lawlor, Leonard. 2020. 'Intelligence and Invention: The Three Aspects of Virtuality in Bergson'. *Interpreting Bergson: Critical Essays*. Ed. Alex Lefebvre and Nils Schott. Cambridge: Cambridge University Press.

Lawlor, Leonard and Valentine Moulard-Leonard. 2016. 'Henri Bergson'. *The Stanford Encyclopedia of Philosophy*. <https://plato.stanford.edu/archives/sum2016/entries/bergson/>.

Le Roy, Charles. 1802. *Lettres philosophiques sur l'intelligence et la perfectibilité des animaux, avec quelques lettres sur l'homme*. Paris: Bossange, Masson et Besson.

Lefebvre, Alexandre. 2008. *The Image of Law: Deleuze, Bergson, Spinoza*. Stanford: Stanford University Press.

Leibniz, G. W. 1969. *Philosophical Papers and Letters*. 2nd ed. Ed. Leroy Loemker. Dordrecht: D. Reidel.

Leibniz, G. W. 1989. *Philosophical Essays*. Trans. Roger Ariew and Daniel Garber. Indianapolis: Hackett.

Levin, D. M. 1993. *Modernity and the Hegemony of Vision*. Berkeley: California University Press.

Levins, Richard, and Richard Lewontin. 1985. 'The Organism as the Subject and Object of Evolution'. *The Dialectical Biologist*. Cambridge: Harvard University Press. 85–106.

Levit, Georgy and Lennart Olsson. 2006. '"Evolution on Rails": Mechanisms and Levels of Orthogenesis'. *Annals of the History and Philosophy of Biology*. Vol. 11: 99–138.

Levy, Arnon. 2013. 'Three Kinds of "New Mechanism"'. *Biology and Philosophy*. Vol. 28: 99–114.

Lewontin, Richard. 2000. *The Triple Helix: Gene, Organism, and Environment*. Cambridge: Harvard University Press.

Lorand, Ruth. 1999. 'Bergson's Concept of Art'. *The British Journal of Aesthetics*. Vol. 39, Issue 4: 400–15.

Lowe, Christopher and Gregory Wray. 1997. 'Radical Alterations in the Roles of Homeobox Genes During Echinoderm Evolution'. *Nature*. Vol. 389, Issue 6652: 718–21.

Lowe, Victor. 1949. 'The Influence of Bergson, James, and Alexander on Whitehead'. *Journal of the History of Ideas*. Vol. 10, Issue 2: 267–96.

Lucas, Heather, Paller, Ken, and Joel Voss. 2012. 'On the Pervasive Influences of Implicit Memory'. *Cognitive Neuroscience*. Vol. 3, Issues 3–4: 219–26.

Lundy, Craig. 2018a. *Deleuze's* Bergsonism. Edinburgh: Edinburgh University Press.

Lundy, Craig. 2018b. 'Bergson's Method of Problematisation and the Pursuit of Metaphysical Precision'. *Angelaki*. Vol. 23, Issue 2: 31–44.

Machamer, Peter, Darden, Lindley, and Carl Craver. 2000. 'Thinking about Mechanisms'. *Philosophy of Science*. Issue 67: 1–25.

McKie, Douglas. 1944. 'Wöhler's "Synthetic" Urea and the Rejection of Vitalism: A Chemical Legend'. *Nature* 153: 608–10

McNab, Brian. 1997. 'On the Utility of Uniformity in Definition of Basal Rate of Metabolism'. *Physiological Zoology*. Vol. 70, Issue 6: 718–20.

McNamara, Patrick. 1996. 'Bergson's "Matter and Memory" and Modern Selectionist Theories of Memory'. *Brain Cognition*. Vol. 30, Issue 2: 215–31.

Madelrieux, Stéphane and Nils Schott. 2020. 'Bergson and Naturalism'. *Interpreting Bergson: Critical Essays*. Ed. Alexandre Lefebvre and Nils F. Schott. Cambridge: Cambridge University Press. 48–66.

Maras, Steven. 1998. 'The Bergsonian Model of Actualization'. *SubStance*. Vol. 27, Issue 85: 48–70.

Margulis, Lynn and Dorion Sagan. 2002. *Acquiring Genomes*. New York: Basic Books.

Marrati, Paola. 2010. 'The Natural Cyborg: The Stakes of Bergson's Philosophy of Evolution'. *The Southern Journal of Philosophy*. Vol. 48: 3–17.

Mates, Benson. 1972. 'Individuals and Modality in the Philosophy of Leibniz'. *Studia Leibnitiana* IV: 81–118.

Maturana, Humberto and Francisco Varela. 1980. *Autopoiesis and Cognition: The Realization of the Living*. Dordrecht: D. Reidel.

Maynard Smith, J., Burian, R., Kauffman, S., Alberch, P., Campbell, J., Goodwin, B., Lande, R., Raup, D., and L. Wolpert. 1985. 'Developmental Constraints and Evolution: A Perspective from the Mountain Lake Conference on Development and Evolution'. *The Quarterly Review of Biology*. Vol. 60, Issue 3: 265–87.

Mayr, Ernst. 1982. *The Growth of Biological Thought*. Cambridge: Harvard University Press.

Mengel, Gregory. 2009. 'The Future Is and Is Not the Past: Heredity, Epigenetics, and the Developmental Turn'. *The Evolutionary Epic: Science's Story and Humanity's Response*. Ed. C. Genet. Santa Margarita: Collins Foundation. 213–17.

Menzies, Peter. 2012. 'The Causal Structure of Mechanisms'. *Studies in History and Philosophy of Biological and Biomedical Sciences*. Issue 43: 796–805.

Michaut, L., Flister, S., Neeb, M., White, K., Certa, U., and W. Gehring. 2003. 'Analysis of the Eye Developmental Pathway in *Drosophila* using DNA Microarrays'. *Proceedings of the National Academy of Sciences of the United States of America*. Vol. 100, Issue 7: 4024–9.

Midgley, David. 2011. '*Creative Evolution*: Bergson's Critique of Science and its Reception in the German-Speaking World'. *The Evolution of Literature: Legacies of Darwin in European Cultures*. Ed. Nicholas Saul and Simon James. Amsterdam: Rodopi. 283–97.

Mill, John Stuart. 1882. *A System of Logic, Ratiocinative and Inductive*. New York: Harper Brothers.

Miquel, Paul-Antoine. 1996. *Le problème de la nouveauté dans l'évolution du vivant. De* L'Évolution créatrice *de Bergson à la biologie contemporaine*. Lille: Presses universitaires du Septentrion.

Miquel, Paul-Antoine. 2007a. 'Bergson and Darwin: From an Immanentist to an Emergentist Approach to Evolution'. *SubStance*. Vol. 36, No. 3, Issue 114: 42–56.

Miquel, Paul-Antoine. 2007b. *Bergson ou l'imagination métaphysique*. Paris: Kimé.

Miquel, Paul-Antoine. 2008. 'Une harmonie en arrière'. *Annales bergsoniennes IV: L'Évolution créatrice 1907–2007: épistémologie et métaphysique*. Ed. Frédéric Worms, Anne Fagot-Largeault, and Jean-Luc Marion. Paris: PUF. 133–45.

Miquel, Paul-Antoine. 2010. 'Chapitre III. De la signification de la vie. L'ordre de la nature et la forme de l'intelligence'. L'Évolution créatrice *de Bergson. Études et commentaires*. Ed. Arnaud François. Paris: Vrin. 175–249.

Miquel, Paul-Antoine. 2011. *Le vital, aspects physiques, aspects métaphysiques*. Paris: Kimé.

Miquel, Paul-Antoine. 2012. 'Le fluide bergsonien est-il toujours bienfaisant?' *Bergson*. Ed. Camille Riquier. Paris: Cerf. 337–57.

Miquel, Paul-Antoine. 2014. *Bergson dans le miroir des sciences*. Paris: Kimé.

Miquel, Paul-Antoine. 2015. *Sur le concept de nature*. Paris: Hermann.

Miquel, Paul-Antoine and Su-Young Hwang. 2016. 'From Physical to Biological Individuation'. *Progress in Biophysics and Molecular Biology*. Vol. 122, Issue 1: 51–7.

Mitcham Carl. 1994. *Thinking Through Technology*. Chicago: University of Chicago Press.

Montebello, Pierre. 2012a. 'La question du finalisme chez Bergson'. *Disséminations de* L'Évolution créatrice. Ed. Hisashi Fujita et al. Hildescheim: Olms Verlag.

Montebello, Pierre. 2012b. 'Deleuze lecteur de Bergson: monisme et naturalisme'. *Bergson*. Ed. Camille Riquier. Paris: Cerf. 251–64.

Moore, F. C. T. 1996. *Bergson: Thinking Backwards*. New York: Cambridge University Press.

Moran, R., Zehetleitner, M., Liesefeld, H. R., Müller, H. J., and M. Usher. 2016. 'Serial vs. Parallel Models of Attention in Visual Search: Accounting for Benchmark RT-Distributions'. *Psychonomic Bulletin Review*. Vol. 23, Issue 5: 1300–15.

Morris, Randall. 1991. *Process Philosophy and Political Ideology: The Social and Political Thought of Alfred North Whitehead and Charles Hartshorne*. Albany: SUNY Press.

Morris, Simon Conway. 2003. *Life's Solution: Inevitable Humans in a Lonely Universe*. Cambridge: Cambridge University Press.

Morris, Simon Conway. 2015. *The Runes of Evolution: How the Universe Became Self-Aware*. Templeton Press.

Mullarkey, John. 2004. 'Forget the Virtual: Bergson, Actualism, and the Refraction of Reality'. *Continental Philosophy Review*. Vol. 37, Issue 4: 469–93.

Mullarkey, John. 2007. 'Life, Movement, and the Fabulation of the Event'. *Theory, Culture, & Society*. Vol. 24, Issue 6: 53–70.

Mullarkey, John. 2008. 'Breaking the Circle: *Élan Vital* as Performative Metaphysics'. *Annales bergsoniennes IV: L'Évolution créatrice 1907–2007: épistémologie et métaphysique*. Ed. Frédéric Worms, Anne Fagot-Largeault, and Jean-Luc Marion. Paris: PUF. 591–600.

Mumford, Stephen and Rani Lill Anjum. 2011. *Getting Causes from Powers*. Oxford: Oxford University Press.

Myers, Charles. 1900. 'Vitalism: A Brief Historical and Critical Review'. *Mind*. Vol. 9, Issue 34: 218–33.

Neumann, C. J., and C. Nüsslein-Volhard. 2000. 'Patterning of the Zebrafish Retina by a Wave of *Sonic Hedgehog* Activity'. *Science*. Vol. 289, Issue 5487: 2137–9.

Ng, S.-F., Lin, R., Laybutt, R., Barres, R., Owens, J., and M. Morris. 2010. 'Chronic High-fat Diet in Fathers Programs β-cell Dysfunction in Female Rate Offspring'. *Nature*. Vol. 467, Issue 7318: 963–6.

Nicholson, Daniel. 2012. 'The Concept of Mechanism in Biology'. *Studies in History and Philosophy of Biological and Biomedical Sciences*. Vol. 43, Issue 1: 152–63.

Noë, Alva. 2004. *Action in Perception*. Cambridge: MIT Press.

Noë, Alva. 2012. *Varieties of Presence*. Cambridge: Harvard University Press.

Nola, Robert and Gürol Irzik. 2005. *Philosophy, Science, Education and Culture*. Dordrecht: Springer.

Normandin, Sebastian and Charles Wolfe. Eds. 2013. *Vitalism and the Scientific Image in Post-Enlightenment Life Science, 1800–2010*. New York: Springer.

Northrop, F. S. C. 1941. 'The Importance of the Bergsonian Influence'. *The Philosophy of Alfred North Whitehead*. Ed. Paul Schilpp. Chicago: Open Court.

Novak, B., and J. J. Tyson. 2008. 'Design Principles of Biochemical Oscillators'. *Nature Reviews Molecular Cell Biology*. Vol. 9, Issue 12: 981–91.

O'Callaghan, Casey. 2013. 'Hearing, Philosophical Perspectives'. *Encyclopedia of the Mind*. Ed. Harold Pashler. SAGE 8: 388–90.

O'Donnell, S., Bulova, S., DeLeon, S., Khodak, P., Miller, S., and E. Sulger. 2015. 'Distributed Cognition and Social Brains: Reductions in Mushroom Body Investment Accompanied the Origins of Sociality in Wasps (Hymenoptera: Vespidae)'. *Proceedings of the Royal Society B: Biological Sciences*. Vol. 282, Issue 1810: 20150791.

Odling-Smee, John, Laland, Kevin and Marcus Feldman. 2003. *Niche Construction: The Neglected Process in Evolution*. Princeton: Princeton University Press.

Ogura, Atsushi, Kazuho, Ikeo, and Takashi Gojobori. 2004. 'Comparative Analysis of Gene Expression for Convergent Evolution of Camera Eye between Octopus and Human'. *Genome Research*. Vol. 14, Issue 8: 1555–61.

Otis, Laura. 1994. *Organic Memory: History and the Body in the Late Nineteenth and Early Twentieth Centuries*. Lincoln: University of Nebraska Press.

Oyama, Susan. 2000. *The Ontogeny of Information: Developmental Systems and Evolution*. 2nd ed. Durham: Duke University Press.

Oyama, Susan, Griffiths, Paul and Russell Gray. 2001. 'Introduction: What is Developmental Systems Theory?' *Cycles of Contingency: Developmental Systems and Evolution*. Ed. Susan Oyama, Paul Griffiths and Russell Gray. Cambridge: MIT. 1–11.

Panero, Alain. 2012. 'Y a-t-il une contingence de la durée chez Bergson?' *Bergson*. Ed. Camille Riquier. Paris: Cerf. 423–35.

Parker, G. A. and John Maynard Smith. 1990. 'Optimality Theory in Evolutionary Biology'. *Nature*. Vol. 348, Issue 6296: 27–33.

Parmentier, Marc. 2017. 'Virtualité et théorie de la perception chez Bergson'. *Methods: savoirs et textes* 17. <https://journals.openedition.org/methodos/4685?lang=en>

Pashler, Harold. 1998. *The Psychology of Attention*. Cambridge: MIT.

Paul, Diane. 1988. 'The Selection of the "Survival of the Fittest"'. *Journal of the History of Biology*. Vol. 21, Issue 3: 411–24.

Pearce, Trevor. 2011. 'Evolution and Constraints on Variation: Variant Specification and Range of Assessment'. *Philosophy of Science*. Vol. 78: 739–51.

Pearce, Trevor. 2012. 'Convergence and Parallelism in Evolution: A Neo-Gouldian Account'. *British Journal for the Philosophy of Science.* Vol. 63, Issue 2: 429–48.

Perri, Trevor. 2014. 'Bergson's Philosophy of Memory'. *Philosophy Compass.* Vol. 9, Issue 12: 837–47.

Peterson, Erik. 2011. 'The Excluded Philosophy of Evo-Devo? Revisiting C. H. Waddington's Failed Attempt to Embed Alfred North Whitehead's "Organicism" in Evolutionary Biology'. *History and Philosophy of the Life Sciences.* Issue 33: 301–20.

Pilkington, A. E. 1976. *Bergson and His Influence: A Reassessment.* New York: Cambridge University Press.

Pineda, D., Gonzalez, J., Callaerts, P., Ikeo, K., Gehring, W. and E. Salo. 2000. 'Searching for the Prototypic Eye Genetic Network: *Sine oculis* is Essential for Eye Regeneration in Planarians'. *Proceedings of the National Academy of Sciences.* Vol. 97, Issue 9: 4525–9.

Pöppel, Ernst. 1978. 'Time Perception'. *Handbook of Sensory Physiology, Vol. 8: Perception.* Berlin: Springer. 713–29.

Post, David and Eric Palkovacs. 2009. 'Eco-evolutionary Feedbacks in Community and Ecosystem Ecology: Interactions Between the Ecological Theatre and the Evolutionary Play'. *Philosophical Transactions of the Royal Society of London B: Biological Sciences.* Vol. 264, Issue 1523: 1629–40.

Posteraro, Tano. 2014. 'On the Utility of Virtuality for Relating Abilities and Affordances'. *Ecological Psychology.* Vol. 26, Issue 4: 353–67.

Posteraro, Tano. 2015. 'Deleuze's Larval Subject and the Question of Bodily Time'. *Symposium: Canadian Journal of Continental Philosophy.* Vol. 19, Issue 2: 187–211.

Posteraro, Tano. 2021a. 'Vitalism and the Problem of Individuation: Another Look at Bergson's *Élan Vital*'. *Vitalism and its Legacies in the 20th Century.* Ed. Charles Wolfe and Chris Donohue. New York: Springer.

Posteraro, Tano. 2021b. 'The Psychological Interpretation of Life'. *The Bergsonian Mind.* Ed. Mark Sinclair and Yaron Wolf. New York: Routledge.

Powell, Russell. 2007. 'Is Convergence More Than an Analogy? Homoplasy and its Implications for Macroevolutionary Predictability'. *Biology & Philosophy.* Vol. 22, Issue 4: 565–78.

Powell, Russell. 2009. 'Contingency and Convergence in Macroevolution: A Reply to John Beatty'. *Journal of Philosophy.* Vol. 106, Issue 7: 390–403.

Pradeu, Thomas and Edgardo Carosella. 2006. 'The Self Model and the Conception of Biological Identity in Immunology'. *Biology and Philosophy.* Vol. 21, Issue 2: 235–52.

Prelorentzos, Iannis. 2008. 'Le problème de la délimitation des choses, des qualités et des états dans la continuité du tout de la réalite selon Bergson'. *Annales bergsoniennes IV: L'Évolution créatrice 1907–2007: épistémologie et métaphysique.* Ed. Frédéric Worms, Anne Fagot-Largeault, and Jean-Luc Marion. Paris: PUF. 433–66.

Prigogine, Ilya and Isabelle Stengers. 1985. *Order Out of Chaos: Man's New Dialogue with Nature.* London: Flamingo.

Prochiantz, Alain. 2008. 'La forme n'est qu'un instantané pris sur une transition'. *Annales bergsoniennes IV: L'Évolution créatrice 1907–2007: épistémologie et métaphysique.* Ed. Frédéric Worms, Anne Fagot-Largeault, and Jean-Luc Marion. Paris: PUF 201–11.

Protevi, John. 2013. *Life, War, Earth: Deleuze and the Sciences.* Minneapolis: University of Minnesota Press.

Quick, Tom. 2017. 'Disciplining Physiological Psychology: Cinematographs as Epistemic Devices in the Work of Henri Bergson and Charles Scott Sherrington'. *Science in Context.* Vol. 30, Issue 4: 423–74.

Ramberg, Peter. 2000. 'The Death of Vitalism and the Birth of Organic Chemistry'. *Ambix.* Vol. 47, Issue 3: 170–95.

Ray, S. and A. B. Reddy. 2016. 'Cross-talk Between Circadian Clocks, Sleep-Wake Cycles, and Metabolic Networks: Dispelling the Darkness'. *BioEssays.* Vol. 38: 394–405.

Reill, Peter Hans. 2005. *Vitalizing Nature in the Enlightenment.* Berkeley: University of California Press.

Reneker, J., Lyons, E., Conant, G. C., Pires, J. C., Freeling, M., Shyu, C. R., and D. Korkin. 2012. 'Long Identical Multispecies Elements in Plant and Animal Genomes'. *Proceedings of the National Academy of Sciences.* Vol. 109, Issue 19: 1183–91.

Ricqlès, Armand de. 2008. 'Cent ans après: *L'Évolution créatrice* au péril de l'évolutionnisme contemporain'. *Annales bergsoniennes IV: L'Évolution créatrice 1907–2007: épistémologie et métaphysique.* Ed. Frédéric Worms, Anne Fagot-Largeault, and Jean-Luc Marion. Paris: PUF. 111–32.

Riquier, Camille. 2008. 'Causalité et création: l'élan vital contre Plotin et la cause emanative'. *Annales bergsoniennes IV: L'Évolution créatrice 1907–2007: épistémologie et métaphysique.* Ed. Frédéric Worms, Anne Fagot-Largeault, and Jean-Luc Marion. Paris: PUF 293–305.

Riquier, Camille. 2010. 'Chapitre II. Les directions divergentes de l'évolution de la vie. Torpeur, intelligence, instinct'. *L'Évolution créatrice de Bergson. Études et commentaires.* Ed. Arnaud François. Paris: Vrin. 125–66.

Robinson, Keith. 2009. 'Introduction: Deleuze, Whitehead, Bergson – Rhizomatic Connections'. *Deleuze, Whitehead, Bergson: Rhizomatic Connections.* Ed. Keith Robinson. New York: Palgrave Macmillan. 1–27.

Roe, Shirley. 1981. *Matter, Life and Generation: Eighteenth-Century Embryology and the Haller–Wolff Debate*. Cambridge: Cambridge University Press.

Roffe, Jon. 2019. 'The Egg: Deleuze Between Darwin and Ruyer'. *Deleuze and Evolutionary Theory*. Ed. Michael Bennett and Tano Posteraro. Edinburgh: Edinburgh University Press. 42–58.

Rosenberg, Alex. 2002. 'Roundtable Discussion 1'. *Promises and Limits of Reductionism in the Biomedical Sciences*. Ed. Marc van Regenmortel and David Hull. Hoboken: John Wiley & Sons. 113–24.

Rosenberg, Eugene and Ilana Zilber-Rosenberg. 2016. 'Microbes Drive Evolution of Animals and Plants: the Hologenome Concept'. *American Society for Microbiology*. Vol. 7, Issue 2: e01395–15.

Roubenoff, Ronnen. 2003. 'Catabolism of Aging: Is it an Inflammatory Process?' *Current Opinion in Clinical Nutrition and Metabolic Care*. Vol. 6, Issue 3: 295–9.

Roughgarden, J., Gilbert, S., Rosenberg, E., Zilber-Rosenberg, I., and E. Lloyd. 2018. 'Holobionts as Units of Selection and a Model of Their Population Dynamics and Evolution'. *Biological Theory*. Vol. 13, Issue 1: 44–65.

Ruse, Michael. 1996. *Monad to Man: The Concept of Progress in Evolutionary Biology*. Cambridge: Harvard University Press.

Ruse, Scott. 2002. 'The Critique of Intellect: Henri Bergson's Prologue to an Organic Epistemology'. *Continental Philosophy Review*. Vol. 35: 281–302.

Ruse, M. Scott. 2005. 'Technology and the Evolution of the Human: From Bergson to the Philosophy of Technology'. *Essays in Philosophy*. Vol. 6, Issue 1: 213–25.

Ruyer, Raymond. 2016. *Neofinalism*. Trans. Alyosha Edlebi. Minneapolis: University of Minnesota Press.

Ruyer, Raymond. 2019. 'Bergson and the Ammophila Sphex'. Trans. Tano Posteraro. *Angelaki*. Vol. 24, Issue 5: 134–47.

Rynasiewicz, Robert. 1995. 'By Their Properties, Causes and Effects: Newton's Scholium on Time, Space, Place and Motion. Part I: The Text'. *Studies in History and Philosophy of Science*. Vol. 26: 133–53.

Sagan, Lynn. 1967. 'On the Origin of Mitosing Cells'. *Journal of Theoretical Biology*. Vol. 14, Issue 3: 225–74.

Salazar-Cuidad, Isaac. 2007. 'On the Origins of Morphological Variation, Canalization, Robustness, and Evolvability'. *Integrative and Comparative Biology*. Vol. 47, Issue 3: 390–400.

Sansom, Roger. 2013. 'Constraining the Adaptationism Debate'. *Biology and Philosophy*. Vol. 13, Issue 4: 493–512.

Sapp, Jan. 2003. *Genesis: The Evolution of Biology*. New York: Oxford University Press.

Scharfstein, Ben-Ami. 1942. *Roots of Bergson's Philosophy*. New York: Columbia University Press.

Schechner, Richard. 2013. *Performance Studies: An Introduction*. 3rd ed. Ed. Sara Brady. New York: Routledge.

Schelling, F. J. W. 1988. *Ideas for a Philosophy of Nature*. Trans. E. E. Harris and P. Heath. Cambridge: Cambridge University Press.

Schelling, F. J. W. 2000. *The Ages of the World*. Trans. Jason Wirth. Albany: SUNY Press.

Schrödinger, Edwin. 1944. *What is Life? Mind and Matter*. Cambridge: Cambridge University Press.

Serb, Jeanne and Douglas Eernisse. 2008. 'Charting Evolution's Trajectory: Using Molluscan Eye Diversity to Understand Parallel and Convergent Evolution'. *Evolution: Education and Outreach*. Vol. 1, Issue 4: 439–47.

Shermer, Michael. 2006. *Why Darwin Matters: The Case Against Intelligent Design*. New York: Henry Holt and Company.

Shields, Christopher. 1986. 'Leibniz's Doctrine of the Striving Possibles'. *Journal of the History of Philosophy*. Vol. 24, Issue 3: 343–57.

Shubin, N. H., Tabin, C. and S. B. Carroll. 2009. 'Deep Homology and the Origins of Evolutionary Novelty'. *Nature*. Vol. 457, Issue 7231: 818–23.

Sinclair, Mark. 2015. 'Is there a Dispositional Modality? Maine de Biran and Ravaisson on Agency and Inclination'. *History of Philosophy Quarterly*. Vol. 32, Issue 2: 161–79.

Sinclair, Mark. 2019. *Being Inclined: Félix Ravaisson's Philosophy of Habit*. New York: Oxford University Press.

Sinclair, Mark. 2020. *Bergson*. New York: Routledge.

Skipper, Robert and Roberta Millstein. 2005. 'Thinking about Evolutionary Mechanisms: Natural Selection'. *Studies in History and Philosophy of Biological and Biomedical Sciences*. Vol. 36, Issue 2: 327–47.

Skogh, C., Garm, A., Nilsson, D. E., and P. Ekström. 2006. 'Bilaterally Symmetrical Rhopalial Nervous System of the Box Jellyfish Tripedalia Cystophora'. *Journal of Morphology*. Vol. 267, Issue 12: 1391–1405.

Smith, Christine and Larry Squire. 2009. 'Medial Temporal Lobe Activity during Retrieval of Semantic Memory Is Related to the Age of the Memory'. *Journal of Neuroscience*. Vol. 29, Issue 4: 930–8.

Smith, Justin 2011. *Divine Machines: Leibniz and the Sciences of Life*. New Jersey: Princeton University Press.

Smith, Olav Bryant. 2004. *Myths of the Self: Narrative Identity and Postmodern Metaphysics*. Lanham: Lexington Books.

Spassov, Spas. 1998. 'Metaphysics and Vitalism in Henri Bergson's Biophilosophy'. *Phenomenology of Life and the Human Creative Condition*. Ed. Anna-Teresa Tymieniecka. Dordrecht: Springer.

Spemann, Hans and Hilde Mangold. 2001. 'Induction of Embryonic Primordia by Implantation of Organizers from a Different Species'. *The International Journal of Developmental Biology*. Vol. 45: 13–38.

Steinert, Steffen. 2017. 'Technology is a Laughing Matter: Bergson, the Comic, and Technology'. *AI & Society*. Vol 32, Issue 2: 201–8.

Stephens, D. W. and J. R. Krebs. 1986. *Foraging Theory*. New Jersey: Princeton University Press.

Sturtevant, Alfred. 1932. 'The Use of Mosaics in the Study of the Developmental Effects of Genes'. *Proceedings of the Sixth International Congregation of Genetics*. Ed. Donald Jones. New York: Macmillan.

Sumner, Francis. 1916. '*The History and Theory of Vitalism* by Hans Driesch; *The Philosophy of Biology* by James Johnstone'. Review. *The Journal of Philosophy, Psychology and Scientific Methods*. Vol. 13, Issue 4: 103–9.

Tahar-Malaussena, Mathilde. 2020. 'The History of the Bergsonian Interpretation of Charles Darwin's Theory of Evolution'. *Global Bergsonism Research Project Webinar*. Department of Philosophy, Penn State University. November 2020.

Taine, Hippolyte. 1871. *On Intelligence*. Trans. T. D. Haye. London: L. Reeve and Co.

Talcott, Samuel. 2019. *Georges Canguilhem and the Problem of Error*. New York: Palgrave Macmillan.

Tauber, Alfred. 2017. *Immunity: The Evolution of an Idea*. New York: Oxford University Press.

Taylor, Timothy. 2012. *The Artificial Ape: How Technology Changed the Course of Human Evolution*. New York: St. Martin's Press.

Tellier, Dimitri. 2008. 'Telle est ma vie intérieure telle est aussi la vie en général'. *Annales bergsoniennes IV: L'Évolution créatrice 1907–2007: épistémologie et métaphysique*. Ed. Frédéric Worms, Anne Fagot-Largeault, and Jean-Luc Marion. Paris: PUF. 423–32.

Terman, Alexei and U. T. Brunk. 2004. 'Aging as a Catabolic Malfunction'. *The International Journal of Biochemistry & Cell Biology*. Vol. 36, Issue 12: 2365–75.

Terman, Alexei. 2006. 'Catabolic Insufficiency and Aging'. *Annals of the New York Academy of Sciences*. Issue 1067: 27–36.

Tomarev, S., Collaerts, P., Kos, L., Zinovieva, R., Halder, G., Gehring, W. and J. Piatigorsky. 1997. 'Squid *Pax-6* and Eye Development'. *Proceedings of the National Academy of Sciences*. Vol. 94, Issue 6: 2421–6.

Tonomura, Akira. 1998. *The Quantum World Unveiled by Electron Waves*. New Jersey: World Scientific.

Trabichet, Luc. 2012. 'Fluidité de la mémoire. La théorie bergsoni-enne de la mémoire face au problème de la nouveauté'. *Bergson*. Ed. Camille Riquier. Paris: Cerf. 437–48.

Turvey, M. 1992. 'Affordances and Prospective Control: An Outline of the Ontology'. *Ecological Psychology*. Vol. 4: 173–87.

Urban, Wilbur. 1951. 'Whitehead's Philosophy of Language and Its Relation to His Metaphysics'. *The Philosophy of Alfred North White-head*. Ed. Paul Schilpp. Chicago: Open Court. 301–28.

Van der Veldt, James. 1943. 'The Evolution and Classification of Philo-sophical Life Theories'. *Franciscan Studies*. Vol. 3: 277–305.

Veitch, J. and S. McColl. 1995. 'Modulation of Fluorescent Light: Flicker Rate and Light Source Effects on Visual Performance and Visual Comfort'. *Lighting Research and Technology*. Vol. 27, Issue 4: 243–56.

Vieillard-Baron, Jean-Louis. 2008. 'Bergson et l'idée de loi scientifique'. *Annales bergsoniennes IV: L'Évolution créatrice 1907–2007: épistémolo-gie et métaphysique*. Ed. Frédéric Worms, Anne Fagot-Largeault, and Jean-Luc Marion. Paris: PUF. 575–89.

Viljanen, Valtteri. 2007. 'Field Metaphysic, Power, and Individuation in Spinoza'. *Canadian Journal of Philosophy*. Vol. 37, Issue 3: 393–418.

Villani, Arnauld. 1999. *La gûepe et l'orchidée. Essai sur Gilles Deleuze*. Paris: Belin.

Vollet, Matthias. 2008. 'La vitalisation de la tendance: de Leibniz à Bergson'. *Annales bergsoniennes IV: L'Évolution créatrice 1907–2007: épistémologie et métaphysique*. Ed. Frédéric Worms, Anne Fagot-Largeault, and Jean-Luc Marion. Paris: PUF. 285–92.

Vollet, Matthias. 2012. 'Creativité comme tendancialité'. *Bergson*. Ed. Camille Riquier. Paris: Cerf. 359–73.

Voss, Joel and Donna Bridge. 2014. 'Hippocampal Binding of Novel Information with Dominant Memory Traces Can Support Both Memory Stability and Change'. *Journal of Neuroscience*. Vol. 34, Issue 6: 2203–13.

Voss, Joel and Donna Bridge. 2015. 'Binding Among Select Episodic Elements is Altered via Active Short-term Retrieval'. *Learning and Memory*. Vol. 22, Issue 8: 360–3.

Voss, J., Bridge, D., Cohen, N. and J. Walker. 2017. 'A Closer Look at the Hippocampus and Memory'. *Trends in Cognitive Science*. Vol. 21, Issue 8: 577–88.

Vucinich, Alexander. 1988. *Darwin in Russian Thought*. Berkeley: University of California Press.

Waddington, C. H. 1941. 'Canalization of Development and the Inher-itance of Acquired Characters'. *Nature*. Vol. 150, Issue 3811: 563–5.

Waddington, C. H. 1956. *Principles of Embryology*. New York: Macmillan.

Waddington, C. H. 1957. *The Strategy of the Genes: A Discussion of Some Aspects of Theoretical Biology*. London: George Allen and Unwin.

Wagner, Gunter and Hans Larsson. 2007. 'Fins and Limbs in the Study of Evolutionary Novelties'. *Fins into Limbs: Evolution, Development, and Transformation*. Ed. Brian Hall. Chicago: University of Chicago Press. 49–61.

Walther, Claudia and Peter Gruss. 1991. '*Pax*-6, a Murine Paired Box Gene, is Expressed in the Developing CNS'. *Development* 113: 1435–49.

Weismann, August. 1891. 'The Continuity of the Germ-Plasm as the Foundation of a Theory of Heredity'. Trans. Selmar Schönland. *Essays on Heredity and Kindred Biological Problems*, Vol. 1. Ed. Edward Poulton, Selmar Schönland, and Arthur Shipley. London: Oxford University Press. 163–256.

Weismann, August. 1893. *The Germ-Plasm: A Theory of Heredity*. Trans. W. Newton Parker and Harriet Rönnfeldt. London: Walter Scott.

Weiss, Paul. 1952. 'The Perception of Stars'. *The Review of Metaphysics*. Vol. 6, Issue 2: 233–8.

Wesson, Robert. 1991. *Beyond Natural Selection*. Cambridge: MIT.

West-Eberhard, Mary Jane. 2003. *Developmental Plasticity and Evolution*. New York: Oxford University Press.

Whitehead, Alfred North. 1922. *The Principle of Relativity with Applications to Physical Science*. Cambridge: Cambridge University Press.

Whitehead, Alfred North. 1964. *The Concept of Nature*. Cambridge: Cambridge University Press.

Whitehead, Alfred North. 1967. *Science and the Modern World*. New York: Free Press.

Whitehead, Alfred North. 1978. *Process and Reality: An Essay in Cosmology*. Ed. David Ray Griffin and Donald Sherburne. New York: Free Press.

Wilkins, Adam. 1997. 'Canalization: A Molecular Genetic Perspective'. *BioEssays*. Vol. 19, Issue 3: 257–62.

Williams, E. A. 2003. *A Cultural History of Medical Vitalism in Enlightenment Montpellier*. Aldershot: Ashgate.

Willmer, Pat. 2003. 'Convergence and Homoplasy in the Evolution of Organismal Form'. *Origination of Organismal Form: Beyond the Gene in Developmental and Evolutionary Biology*. Ed. Gerd Müller and Stuart Newman. Cambridge: MIT. 33–50.

Wilson, P., Wolfe, A., Armbruster, S. and J. Thompson. 2007. 'Constrained Lability in Floral Evolution: Counting Convergent Origins of Hummingbird Pollination in *Penstemon* and *Keckiella*'. *New Phytologist*. Vol. 176, Issue 4: 883–90.

Wimber, M., Alink, A., Charest, I., Kriegeskorte, N. and M. Anderson. 2015. 'Retrieval Induces Adaptive Forgetting of Competing Memories via Cortical Pattern Suppression'. *Nature Neuroscience*. Vol. 18, Issue 4: 582–9.

Winther, Rasmus Grønfeldt. 2006. 'Parts and Theories in Compositional Biology'. *Biology and Philosophy*. Vol. 21: 471–99.

Wolfe, Charles. 2011. 'From Substantival to Functional Vitalism and Beyond: Animas, Organisms and Attitudes'. *Eidos*. No. 14: 212–35.

Wolfe, Charles. 2014. 'The Organism as Ontological Go-Between: Hybridity, Boundaries and Degrees of Reality in its Conceptual History'. *Studies in History and Philosophy of Biological and Biomedical Sciences*. Issue 48: 151–61.

Wolfe, Charles. 2017. 'Materialism New and Old'. *Antropología Experimental*. Vol. 17, Issue 13: 215–24.

Wolfe, Charles and Andy Wong. 2014. 'The Return of Vitalism: Canguilhem, Bergson, and the Project of Biophilosophy'. *The Care of Life: Transdisciplinary Perspectives in Bioethics and Biopolitics*. Ed. Miguel de Beistegui, Giuseppe Bianco, and Marjorie Gracieuse. New York: Rowman & Littlefield. 63–75.

Wolfe, Charles and M. Terada. 2008. 'The Animal Economy and Object and Program in Montpellier Vitalism'. *Science in Context*. Vol. 21, Issue 4: 537–79.

Wolsky, Maria de Issekutz and Alexander Wolsky. 1992. 'Bergson's Vitalism and Modern Biology'. *The Crisis in Modernism: The Vitalist Controversy*. Ed. Frederick Burwick and Paul Douglass. Cambridge: Cambridge University Press.

Wong, Andy Tai Tak. 2016. *Critical Life: Bergson, Canguilhem and the Critical Investigation of Life and the Living* (Doctoral Dissertation). Faculté de Philosophie et Lettres, Département de Philosophie. Université de Liège.

Woodger, Joseph Henry. 1929. *Biological Principles: A Critical Study*. New York: Harcourt.

Worms, Frédéric. 1997. *Introduction à Matière et mémoire de Bergson*. Paris: PUF.

Worms, Frédéric. 2000. *Le vocabulaire de Bergson*. Paris: Ellipses.

Worms, Frédéric. 2004. *Bergson ou Les deux sens de la vie: étude inédite*. Paris: PUF.

Worms, Frédéric. 2008. 'Ce que est vital dans *L'Évolution créatrice*'. *Annales bergsoniennes IV: L'Évolution créatrice 1907–2007: épistémologie et métaphysique*. Ed. Frédéric Worms, Anne Fagot-Largeault, and Jean-Luc Marion. Paris: PUF. 641–52.

Worms, Frédéric. 2010. 'Quel vitalisme au-delà de quell nihilism? De *L'Évolution créatrice* à aujourd'hui'. L'Évolution créatrice *de Bergson. Études et commentaires.* Ed. Arnaud François. Paris: Vrin. 251–9.

Worms, Frédéric, Fagot-Largeault, Anne and Jean-Luc Marion. Eds. 2008. *Annales bergsoniennes IV: L'Évolution Créatrice 1907–2007: épistémologie et métaphysique.* Paris: PUF.

Wouters, Arno. 2005. 'The Function Debate in Philosophy'. *Acta Biotheoretica.* Vol. 53: 123–51.

Wyk, Alan van. 2012. 'What Matters Now?' *Cosmos and History.* Vol. 8, Issue 2: 130–6.

Wyszecki, Günter and W. S. Stiles. 1982. *Colour Science: Concepts and Methods, Quantitative Data and Formulae.* 2nd ed. New York: Wiley.

Yoshida, Masa-aki, Yura, Kei and Atsushi Ogura. 2014. 'Cephalopod Eye Evolution was Modulated by the Acquisition of *Pax*-6 Splicing Variants'. *Nature: Scientific Reports.* Vol. 4, No. 4256. <https://doi. org/10.1038/srep04256>

Yoxen, E. J. 1979. 'Where Does Schrödinger's *What is Life?* Belong in the History of Molecular Biology?' *History of Science.* Vol. 17: 17–52.

Zalts, Harel and Itai Yanai. 2017. 'Developmental Constraints Shape the Evolution of the Nematode Mid-Developmental Transition'. *Nature: Ecology & Evolution.* Vol. 1, No. 0113.

Zemansky, M. W. 1968. *Heat and Thermodynamics*, 5th ed. New York: McGraw-Hill.

Zirkle, Conway. 1946. 'The Early History of the Idea of the Inheritance of Acquired Characters and of Pangenesis'. *Transactions of the American Philosophical Society.* Vol. 35: 91–151.

Zylinksi, Sarah and Sönke Johnsen. 2014. 'Visual Cognition in Deep Sea Cephalopods'. *Cephalopod Cognition.* Ed. Anne-Sophie Darmaillacq, Ludovic Dickel, and Jennifer Mather. Cambridge: Cambridge University Press. 223–41.

Index